DÉTENTE AND BEYOND

ANGLO-ROMANIAN RELATIONS IN THE AVIATION INDUSTRY (1966-1993)

P.I.E. Peter Lang
Bruxelles · Bern · Berlin · New York · Oxford · Wien

EUROCLIO is a scientific and editorial project, a network of research institutions and researchers, and an ideas forum. EUROCLIO as an editorial project consists of two aspects: the first concerns studies and documents, the second concerns tools. Both are aimed at making the results of research more accessible, and also at opening up paths through the history of European construction/integration/unification.

The EUROCLIO series meets a dual objective:

- to provide reference tools for research,
- to provide a platform for this research in terms of the publication of results.

The series thus consists of two sub-series that satisfy these requirements: the Studies and Documents series and the References series. These two series are aimed at general libraries and/or university history departments, teachers and researchers, and in certain cases, specific professional circles.

The Studies and Documents series consists of monographs, collections of articles, conference proceedings, and collections of texts with notes for teaching purposes.

The References series consists of bibliographies, guides and other tools. It thus contributes to the creation of a database making up a "Permanent catalogue of sources and bibliographies on European construction".

DÉTENTE AND BEYOND

ANGLO-ROMANIAN RELATIONS IN THE AVIATION INDUSTRY (1966-1993)

Mauro Elli

Euroclio
Vol. 102

Linguistic supervision: Nicola Resnati.

This volume is published thanks to the support of Università degli Studi di Milano and Università degli Studi di Padova.

This publication has been peer reviewed.

Éditions scientifiques internationales
Brussels, 2018
1 avenue Maurice, B-1050 Brussels, Belgium
brussels@peterlang.com; www.peterlang.com

Printed in Germany

ISSN 0944-2294
ISBN 978-2-8076-0674-6
ePUB 978-2-8076-0676-0
D/2018/5678/20
DOI 10.3726/b13344
ePDF 978-2-8076-0675-3
MOBI 978-2-8076-0677-7

CIP available at the Library of Congress and the British Library.

Bibliographic information published by "Die Deutsche Nationalbibliothek".

"Die Deutsche National Bibliothek" lists this publication in the "Deutsche Nationalbibliografie"; detailed bibliographic data is available on the Internet at <http://dnb.de>.

Table of Contents

Acronyms

AA	Associated Aerospace
AEI	Associated Electrical Industries
AMEJ	Papers of Julian Amery, Churchill College, Cambridge
ANR, CC	The National Archives of Romania, Central Committee of the Romanian Communist Party, Bucharest
ARB	Air Registration Board
BA	British Airways
BAC	British Aircraft Corporation
BAe	British Aerospace
BDOHP	British Diplomatic Oral History Programme, Churchill College, Cambridge
BEA	British European Airways
BHC	British Hovercraft Corporation
B-N	Britten-Norman
BP	British Petroleum
CAA	Civil Aviation Administration
CIAR	Headquarters of the Romanian Aeronautical Industry (Centrala Industriei Aeronautice Române)
CMEA	Council for Mutual Economic Assistance
CNIAR	National Centre of Aeronautical Industry (Centrul Național al Industriei Aeronautice Române)
CoCom	Coordinating Committee for Multilateral Export Controls
CPSU	Communist Party of the Soviet Union
CSCE	Conference on Security and Co-operation in Europe
DBPO	Documents on British Foreign Policy Overseas
DOI	Department of Industry
DOT	Department of Trade
DTI	Department of Trade and Industry

EBRD	European Bank for Reconstruction and Development
ECGD	Export Credits Guarantee Department
EMBRAER	Brazilian Aeronautics Company (Empresa Brasileira de Aeronáutica)
ERC	Exporters Refinance Corporation
FAA	Federal Aviation Administration
FCO	Foreign & Commonwealth Office
FOI/DTI	Papers obtained by the UK Department for Business, Innovation & Skills pursuant the Freedom of Information Act
FOI/FCO	Papers obtained by the UK FCO pursuant the Freedom of Information Act
FOIA/DoS	Papers obtained from the U.S. Department of State pursuant the Freedom of Information Act
GAB	Bucharest Aeronautical Group (Grupul Aeronautic Bucureşti)
HAEC	Historical Archives of the European Communities
HSA	Hawker Siddeley Aviation
IAR	Romanian Aeronautical Industry (Industria Aeronăutică Română)
IATA	International Air Transport Association
IAvB	Bucharest Aircraft Company (Întreprinderea de Avioane Bucureşti)
ICI	Imperial Chemical Industries
ICPAS	Institute for Aerospace Design and Construction (Institutul de Cercetări și Proiectări Aerospațiale)
ILS	instrumental landing system
IMFCA	Institute for Fluid Mechanics and Aerospace Constructions (Institutul de Mecanica Fluidelor şi Construcții Aerospațiale)
INCREST	National Institute for Scientific and Technical Creation (Institutul Național pentru Creație Științifică și Tehnică)
IRMA	Enterprise for the Repair of Aeronautical Material (Întreprinderea de Reparații Material Aeronautic)
LAR	Romanian Airways (Liniile Aeriene Române)
MAEF	Archives du Ministère des affaires étrangères

MAER	Romanian Ministry of Foreign Affairs
MBB	Messerschmitt-Bölkow-Blöhm
Mintech	Ministry of Technology
MoD	Ministry of Defence
NARA, ADD	National Archives and Records Administration, access to archival databases
NSF	National Salvation Front
OECD	Organisation for Economic Co-operation and Development
OEEC	Organisation for European Economic Co-operation
PCL	Polytechnic of Central London
PMR	Romanian Workers' Party (Partidul Muncitoresc Român)
RCP	Romanian Communist Party (Partidul Comunist Român)
RMFA	Archive of the Romanian Ministry of Foreign Affairs
SBAC	Society of British Aircraft Constructors
SNECMA	National Society of Research and Construction of Aviation Engines (Société nationale d'étude et de construction de moteurs d'aviation)
STOL	stort take-off and landing
Tarom	Romanian Air Transport (Societatea de Transporturi Aeriene Române)
TARS	Romanian-Soviet Air Transport (Societatea de Transporturi Aeriene Româno-Sovietice)
UKNA	The National Archives of the United Kingdom
VFW	United Flight-technical Company (Vereinigte Flugtechnische Werke)

Introduction and Acknowledgements

This book has come together almost by chance. I was mainly concerned – and I still am – with nuclear power and in 2010 I was completing a research on one of the two atomic power plants that the British managed to export in the 1950s, specifically the one purchased by the ENI Group for Latina. I was looking for a new topic, convinced – as I was back then – that my job would always be historical research, a field that possibly had nothing to do with nuclear power, because – it was suggested – it would be appropriate for my academic future to show the ability to be flexible, not to focus on one single theme to be rehashed at any given opportunity.

On the one hand, I wanted to investigate something that had to do with Romania, because in previous years I had been able to develop a special bond with this country for personal reasons. On the other hand, I had a very cursory knowledge of its history, and I had never had dealings with Eastern Europe. It was then that one of my colleagues in Milan, an expert on such topics, with his good-naturedly brazen ways, told me that if I wanted to try and research a significant theme that involved both the Romanians and 'my beloved English' I would have to consider planes.

Aeroplanes, for me, were just a means of transport – and not a very comfortable one. However, I jumped aboard one heading for London and for a couple of weeks I lost myself in that extraordinary place of study still known at that time as the Public Record Office. During that stay, I could not do much, apart from seeing the amount of material available and understanding, though still in an obscure way, the enormous quantity of historical links that such research could offer. Accustomed as I was to take the availability of low-cost carriers on European routes for granted, I still remember my amazement upon learning that one of the first aircraft operated by the newly-formed Ryanair – the Spirit of Dublin – was a ROMBAC 1-11 manufactured in Romania.[1]

Everything would have probably ended here, had I not taken part in the Final ESF Inventing Europe Conference – 4th Tensions of Europe Conference, which was to be held in June 2010 in Sofia – a city I had

1 Robert Church, *The One-Eleven Story*, Tonbridge, Air-Britain Historians, 1996, p. 199-203.

never visited before and that I thought would be an interesting place to spend a few days in the company of my wife to be. On a more professional level, the conference offered a great opportunity to use my material on Anglo-Romanian aeronautical relations, but I needed someone who was familiar with the history of aeronautics and who was available to present a paper with me, dealing with the background issues. I turned to David Burigana, whom I already knew, though not particularly well: that was the beginning of an important friendship and the moment when I decided to take seriously the research that culminates in the present volume.

It is not necessary to bring post-Popperian epistemology into the picture as regards the driving forces of scientific research. Rather, it might be useful to make a different type of clarification. Against the backdrop of historical disciplines, where a persistent and perhaps excessive specialisation is only partially balanced by an interdisciplinary approach, which in turn becomes something of a fetish, a codeword to secure funds for research, this book takes on a different dimension. Indeed, it is not a work on the history of the Cold War, nor on economic history, nor on the history of Eastern Europe, but rather a bit of all these things, blended together in varying measure. The boundary between eclecticism and amateurism is somewhat blurred and I can only assure that I have done my best to opt for the former. What I wanted to do was investigate a concrete case of those relationships that crossed the Iron Curtain in Détente Europe. Although the framework of diplomatic relations was already clear, there were very few studies that researched what had happened at the lower levels, especially in terms of economic and technological relations. In order for these relationships to make sense historically and for them to make a certain contribution to their progress, they needed a contextualisation that gleaned from different areas – and, for the most part, they needed to be investigated as a whole.

This is why this book is, in a sense, a kind of 'contamination' between these different areas and it is why I was prompted, at the writing stage, to adopt a concentric-circle structure in the various chapters, recalling the political and economic context of the United Kingdom and Romania, moving on to bilateral relations, and finally addressing aeronautical issues. In addition, I have dwelt on the technical aspects of these issues, because I do not think it is possible to really investigate any historical dynamics that have to do with technical content – whether nuclear or aeronautical – limiting myself to the perceptions of politicians and diplomats, or narrowing down the analysis to a certain type of documentation. I thought, in short, that this way of proceeding, that is, integrating different points of view by anchoring them to a precise bilateral dynamics, was the most functional way to present the heart of the matter, the model that seems to be revealed.

Here another clarification is necessary. I am speaking about a 'model', not about 'conclusions' – and the reader wishing to throw a glance at the table of contents might be surprised at the fact that this book does not offer the much-expected ending chapter entitled 'Conclusions'. I fully realise that this fact can be misleading, but I hasten to assure the reader that I conscientiously considered the matter while planning the book and that I opted for including my conclusions in this introduction. Insofar as a conclusion is not merely a summary of the main points already examined in the body of the text, it should offer an interpretative model, or – even better – a possible historiographical approach for further studies. Therefore, I deemed it useful to offer the reader my key of interpretation beforehand, as well as any proper reference to the general context. Firstly, this would hopefully lead more easily the reader through the narrative of the book – whose chapters are more or less self-contained entities with their own introduction and conclusions. Secondly, the reader can test the model that I suggest hereafter in a straightforward way and he or she is enabled to formulate his or her own ideas more freely, in a more fruitful hermeneutic circle – I believe.

But what model could emerge from such complicated events, lasting nearly three decades? Thirty years seems a short time, but they are in fact quite a long period compared to the 'pulse' of contemporary history: the world, the coordinates of the context in which these events occurred were already radically transformed when it all ended, passing through the transformations of the 1970s and the end of the Cold War. Also, the number of actors in the game, the difficulty of combining political and technical aspects, the apparent breaking down of Anglo-Romanian aeronautical cooperation that, like an underground river, seems to disappear, swallowed up by the earth, only to reappear with new twists. Years ago, still in the early stages of my research and thus with a very partial knowledge of the subject matter, I tried to map these channels into a flowchart. In the end it filled nine pages, held together with adhesive tape. That was probably the first of several times when I thought I would give up this project.

With a certain uneasiness – having to finalise my work – I find it necessary to sketch out its contents, to serve as a key to reading the book, outlining the topics of the single chapters. Thus, it seems to me that a model does emerge. First, at risk of stating the obvious, Anglo-Romanian aeronautical cooperation was based on polarity, not convergence. On one side, the positive pole, the United Kingdom, which possessed technology and manufacturing skills in an important sector of the mechanical industry and was less and less able to compete profitably in the usual markets; on the other side, the negative pole, Romania, which aspired to possess this technology and production skills, but basically could not afford them. Cooperation was not, therefore, an act of 'building together', but rather a 'transfer', which was always conditioned and unidirectional.

Of course, this does not mean that the Romanians were untalented or unable to achieve the same levels of quality as the British. However, the results of the 'building together' strategy were almost negligible, even at a time when it was very fashionable to talk about East-West cooperation – especially in the light of a renewed importance of the European context between the 1960s and the 1970s – as a key synergy to reach new markets in developing countries.[2] Of course, everyone talked about it: Anglo-Romanian cooperation was often seen as a vehicle for selling aeronautical material to African countries, and Romania played an important role in the arrival of Rolls-Royce in China. In the end, however, amidst pretentions and exploitations, nothing came of it.

The whole story could be seen as an example of 'constructive illusion,' or cooperative dynamics based on misperceptions.[3] However, in the face of a recent historiography that has rightly drawn attention back to the cooperative dimension of East-West relations, especially in the European arena, it is fundamental to try and pinpoint the connotations of the concept of 'cooperation.'[4] Undoubtedly, in the present case, the numerous interactions between multiple actors placed on different levels were always brought back to the governmental and diplomatic dimension, because, ultimately, the risk of investment needed to implement Anglo-Romanian cooperation were guaranteed by the state.

The 'transfer' of technology and productive skills was always subject to conditions. It should be noted that such limitations were not only applied

---

2 On the European dimension of Détente during the longer 1970s see Federico Romero and Silvio Pons, "Europe Between the Superpowers, 1968-1981", in Antonio Varsori and Guia Migani (eds.), *Europe in the International Arena During the 1970s*, Brussels, P.I.E. Peter Lang, 2011, p. 85-97; Angela Romano, *From Détente in Europe to European Détente*, Brussels, P.I.E. Peter Lang, 2009; Wilfried Loth and Georges-Henri Soutou, *The Making of Détente. Eastern and Western Europe in the Cold War, 1965-75*, Abingdon, Routledge, 2008.

3 For more on this see Eric Grynaviski, *Constructive Illusions. Misperceiving the Origins of International Cooperation*, Ithaca, Cornell University Press, 2014.

4 See Sari Autio-Sarasmo and Katalin Miklòssy, "The Cold War from a New Perspective", in Sari Autio-Sarasmo and Katalin Miklòssy (eds.), *Reassessing Cold War Europe*, London, Routledge, 2011, p. 1-15.

to cases of military relevance. Of course, as far as the military industry is concerned, Romania's membership in the Warsaw Pact was a very restrictive factor, irrespective of the political positions assumed by Bucharest, especially after the invasion of Czechoslovakia in 1968, which caused the premature end of the contacts initiated about a possible cooperation in the realisation of a supersonic fighter. Conditions and limitations affected the civil sector, for example with regard to certification procedures and the maintenance of design authority by the United Kingdom, partly because it was difficult to control technological developments once the know-how had been transferred, partly because the Romanians never concealed their desire to proceed in both areas – and to consider the military one more important.

The British were also conditioned by relationships with their allies at CoCom. Historians have dealt with multilateral control mechanisms on exports to countries behind the Iron Curtain,[5] notably with respect to the period 1950s-1960s,[6] highlighting the contrasts between the Westerners. In the present case, there was certainly the fear that the control mechanisms could serve to unduly favour American competition and, had there been interests as vital as Rolls-Royce's entry into the Chinese market, the British government would have contemplated the hypothesis of unilateral solutions. However, the overall assessment of CoCom's role remained positive and far from overcoming the limits, in the case of Romania, the British were rather inclined to interpret them restrictively, for example concerning metallurgical techniques or the development of an afterburner for the Viper engine. From a strictly strategic point of view, these elements probably would not have changed the military balance between the two blocs in any way, but why risk tensions with Western allies, and especially with the United States, which was still the most important partner in the aerospace sector?

Moreover, if Anglo-Romanian aeronautical cooperation was indeed based on an intrinsic polarity, the same was true of its 'surrounding conditions.' On the one hand, the United Kingdom was engrossed in the debate on decline and engaged in a policy of differentiation aimed to seize the political and economic opportunities presented by Détente in Europe; on the other

---

5 Frank Cain, *Economic Statecraft During the Cold War. European Responses to the US Trade Embargo*, Oxford, Abington, Routledge, 2013.

6 See, for example, Jari Eloranta and Jari Olaja, *East-West Trade and the Cold War*, Jyväskylä, Jyväskylä University Press, 2005 and Luciano Segreto, East-West Trade in Cold War Europe: National Interests and Hypocrisy, paper presented at the XIV International Economic History Congress, Session 111, Helsinki, August 2006.

hand, Nicolae Ceauşescu's Romania, the maverick of the Soviet Bloc, was registering striking indexes of economic growth.[7] These conditions enabled the activation of a rather dense network of personal and corporate contacts, which, most of the time, represented the prime mover of the various moments in which Anglo-Romanian cooperation was articulated. Naturally, this does not mean that the level of high politics and diplomacy were secondary: it would be structurally impossible in the case of a regime like the one in power in communist Romania, but also in the case of the United Kingdom there is systematic cooperation, although not devoid of tensions, between the diplomatic representation in Bucharest, the departments at Whitehall and the upper echelon of the British government on the one hand, and the industrial subjects with their mediators on the other.

This testimony to the vitality and general competence of a diplomatic tradition, for that matter already highlighted,[8] must be read alongside the fact that rarely did the initiative come through these channels. The most striking case is that of George Pop, for about thirty years the representative and mediator of the interests of the British aeronautical industry in Romania: not only is Pop often seen as the main – sometimes unique – source of information for the Embassy in Bucharest and, consequently, for Whitehall, but through a formidable network of contacts and thanks to an extraordinary knowledge of the power mechanisms of the Romanian regime, he was able, on more than one occasion, to make the difference, suggesting solutions that allowed the continuation of bilateral cooperation.

Examples are numerous and they concern the Romanians as well, ranging from important members of the Communist Nomenklatura, such as Emil Bodnaraş, to experts in matters related to aviation like Ştefan Ispaş. These are cases in which personal sympathies and interests played a significant role in orienting the choices on which bilateral cooperation was set up: elements that are undoubtedly difficult to investigate, but whose documentation has left different traces. In other cases, it is also important to consider a perspective that goes beyond the time limits of the Cold War: it is impossible, for example, not to take into consideration the role played by that extraordinary engineer Henri Coandă in the creation of the British and Colonial Airplane Company in Bristol in the years before the First World War, when one appraises the easiness his contacts with

---

7 For a fresh portrait of Ceauşescu see Eliza Gheorghe, "Nicolae Ceauşescu", in Steven Casey and Jonathan Wright (eds.), *Mental Maps in the Era of Détente and the End of the Cold War 1968-91*, Basingstoke, Palgrave, 2015, p. 60-80.

8 John Young, *Twentieth Century Diplomacy: A Case Study of British Practice, 1963-1976*, Cambridge, Cambridge University Press, 2008.

Rolls-Royce – which had inherited the former company – in the early 1970s. In the end, the constant reminder of Ceauşescu's aviation projects was Romania between the two wars, albeit always accompanied by the ideological view that communism should allow the full realisation of an industry that the old 'aristocratic-bourgeois' regime had not been able to develop.

These personal contacts were activated whenever British companies had any industrial and commercial interest. I should also mention what I consider to be the missing piece of this research, that is, the lack of the documentation produced by these companies, which would have been of paramount importance. I have tried, on several occasions, to contact Rolls-Royce and BAE Systems, only to get evasive or negative answers about the availability, or the existence, of any documentation. So, I had to try to trace this information through the content of the papers I had access to. Undoubtedly, the industrial and commercial interest was stimulated by competition between Western countries in an attempt to enter previously inaccessible markets, to the point that the British considered less dangerous the possibility of Romania purchasing aircraft from the USSR rather than the United States, a more fearsome competitor. Be that as it may, the readiness to do business with Romania became, as time went by, and especially in the second half of the 1970s, a means of seeking financial protection from the government for products still struggling to find a market (as in the case of the 1-11) or without one (for example, the German VFW-Fokker 614, a substantial failure.)

All this leads to the crucial point in the story. Romania did not, of course, have the financial resources necessary to rebuild its aviation industry, so the costs had to be covered through counter-trade and loans with merchant banks, at least until the export of new products, whose technological know-how and production practices had been previously imported, generated, theoretically speaking, sufficient revenue to pay the debts. This was something very similar to what happened with developing countries and, not by chance, a Romanian leitmotiv during these negotiations was the request to obtain – with little success – the same financial conditions offered to the poorest African and Asian countries.

For years, the political and economic situation of Détente in Europe had justified this line of conduct. In fact, Romania was in a powerful negotiating position: on the one hand, it could put forward the need for a favourable treatment that would politically encourage its independence from Moscow – although, in fact, Romania's individual choices in favour of one Western partner or another were not necessarily equivalent to anti-Soviet positions; on the other hand, in addition to exploiting East-West political competition, Romania could benefit from the commercial rivalry between Western countries. In short, despite all the constraints deriving from being a Warsaw Pact country, Romania could try and follow the example of Yugoslavia.[9]

Analysing in detail the long negotiations on the economic and financial aspects gives an opportunity to realise, not without some degree of amazement, how long the Romanian tactic proved successful, even after the conditions that had originally made it possible began to disappear one after another. In the mid-1970s, Détente began a descending parabola, the repressive nature of Ceauşescu's regime was getting increasingly detestable and the consequences of the oil shock became manifest, especially in terms of inflation and economic depression. Yet, the objective difficulties in fixing costs over the medium to long term and the exhausting haggling on the escalation clauses and the terms of the loans continued to be largely resolved in favour of the Romanians until 1979. This is because, as long as Keynesian neo-corporatist criteria remained valid, the British governments of the time were ultimately available to finance and cover the risk of these operations, in order to contain the employment repercussions over a civil airframe industry that was in dire straits; likewise, it continued to be considered normal, all in all, that the state intervened with industrial policy measures in support of those sectors that, rightly or wrongly, were considered crucial from the economic point of view as well as from that of national prestige. So, what could roughly be described for the British as reasons of internal politics were, for the Romanians, the real lasting advantage in these negotiations.

When questioning the rise of post-Keynesian political economy and its effects on the Cold War,[10] the apparently more logical answer that would seem to stem from the history of Anglo-Romanian aeronautical cooperation is that the arrival of Margaret Thatcher entailed exhaustion of those support mechanisms that had guaranteed the operation until that moment. However, that answer would only be partial. In fact, even if Whitehall's loss of interest was tangible, not least because in the various corporate evolutions of the British aerospace sector, those who were historically concerned with the relationship with Romania came progressively to be less influential, it should also be noted that, until the fall of Communism, cooperation projects basically stalled because of

9 See Benedetto Zaccaria, *The EEC's Yugoslav Policy in Cold War Europe, 1968-1980*, Basingstoke, Palgrave, 2016.

10 See Federico Romero, "Cold War Historiography at the Crossroads", *Cold War History*, Vol. 14, No. 4 (2014), p. 685-703.

Ceaușescu's determination not to contract new debts, and not really for the unwillingness of the British to offer credits. The emergence of new economic paradigms was certainly significant, but had a comparatively lower impact on the fate of these vicissitudes in the 1980s than the devastating austerity imposed by Ceaușescu on Romania, in turn a reaction to the global crisis of the previous decade and further confirmation that even the absence of reforms and a 'totalitarian' social control were not a guarantee for the good functioning of planned economies.[11]

All these paradigms were decisive after Ceaușescu's death and the end of the Cold War, when it now seemed a self-evident truth that a liberist shock therapy would be the most effective recipe for post-communist economies. Henceforth the idea that the resources needed for the revival of Romanian aeronautical projects should first come from the restructuring of the Romanian aeronautical industry itself. Only later, this restructuring having been carried out with sacrifices, would other sources intervene and, in short, what could be done by the British was to help the Romanians help themselves. All this signed the death warrant of almost thirty years of cooperation and redefined the role of the Romanian industry in partnership with Boeing.

These being the general conclusions, the model that I believe can be derived from this investigation – and that I hope might be useful in a comparative key for those who will in future be concerned with trans-European economic interactions during the Cold War[12] – can benefit from a brief presentation of the structure of the book, even though the material could perhaps be arranged differently.

Chapter 1 presents the polarity represented by the United Kingdom-Romania binomial, with particular attention to the political, diplomatic, economic and industrial context data and a sight to key historiographical issues: Détente and Romania's 'independence' from the Soviet Union and its relations with Western countries. The chapter analyses the launch of Anglo-Romanian aeronautical cooperation, starting with the sale of five BAC 1-11s to modernise the TAROM fleet and enable it to exploit better tourist flows, a valuable currency source. This transaction and all the related counter-trade measures designed to make it possible opened

---

11 For the opposite view, regarding the USSR, see Philip Hanson, *The Rise and Fall of the Soviet Economy. An Economic History of the USSR from 1945*, New York, Longman, 2003.

12 Angela Romano and Federico Romero, "European Socialist Regimes Facing Globalisation and European Co-operation: Dilemmas and Responses", *European Review of History*, Vol. 21, No. 2 (2014), p. 157-164.

up a series of long-term prospects, not limited to the civilian sector, which seemed to tie Romania steadily to the British industry, giving the latter a new and significant commercial outlet.

Chapter 2 considers the most interesting element of counter-trade, that is, the building under licence of the Britten-Norman Islander in Romania. The Romanian line should not have simply served to generate revenue to cover the cost of the aircraft purchased by TAROM but, above all, to mature production skills that could satisfy Western quality controls, starting with a simple plane and promising commercial prospects. The Islander, at the crossroads of all Anglo-Romanian aeronautical relations during the 1970s, might have actually been a success in terms of developing industrial skills, but Romania did not succeed in autonomously producing even half of the plane components. In addition, the Islander affair clearly unearths the financial, if not managerial, weakness of the British aeronautical industry in the 1970s, even if it could rely on a high-quality product that later enjoyed worldwide business success.

Chapter 3 examines relations in the military sector, stressing the importance of Romania being a Warsaw Pact country after the invasion of Czechoslovakia in 1968, irrespective of the unequivocal condemnation of the incident by Ceauşescu and its convergence with Yugoslavia. With the possibility of cooperating in the construction of a supersonic fighter being denied early on, the Romanian interest shifted onto the ability to build some Rolls-Royce engines under licence, starting with the Viper. This was to become the engine of an aircraft at the heart of Romanian-Yugoslav cooperation and a starting point for repeated attempts to convince Rolls-Royce to collaborate in the design of an afterburner. In general, contacts in the military sector demonstrate, on the one hand, the continued effectiveness of CoCom's restrictions on exchanges with the Soviet bloc and, on the other hand, the impracticability of the ambitions of the regime in the face of the difficulties of the Romanian productive apparatus to assimilate the technology obtained. The relations in this area allowed Rolls-Royce to activate, through good Romanian offices, those contacts with the Chinese authorities that led to the sale of the production licence for a military engine to the People's Republic.

Chapter 4 shows the rapid decline of the prospects for civilian cooperation that had emerged in the late 1960s, due to the growing weakness of the British aviation industry and to Boeing's first foray into the Romanian market. Against this backdrop, starting with Pop's initiative, it was possible to put in place a new transaction involving the reactivation of the 1-11 production line in the United Kingdom thanks to the financial support of the government – be it Conservative or Labourist, convinced as it was that it should intervene in support of an industry operating in areas of the country which had already been heavily affected by unemployment in the 1970s. It was precisely the government's support that would enable the sale at fixed prices, over a period of galloping inflation, of other 1-11s to the Romanians, with a view to setting up a subcontracting production line in Romania as a counter-trade element.

Chapter 5 examines the lengthy negotiations on the ROMBAC project, that is, the production under licence of the 1-11 in Romania in the second half of the 1970s, underlining how the lack of German competition left the British airplane as the only concrete option for the Romanians, as well as the sale of the production licence to Romania was considered a useful means of survival for the British airframe industry. This situation highlights how economical and commercial considerations were decisively predominant over political-diplomatic issues, and how, on part of the British, there was a concern to place precise constraints on the extent of technological transfer and on the degree of Romania's future industrial independence. On the other hand, the frequent variation in the number of planes involved in the project gives a clue to how aleatory Romanian plans were.

Chapter 6 analyses the failure of the ROMBAC project in its original formulation in the changing climate of the 1980s, while taking into consideration the epilogue of Anglo-Romanian aeronautical cooperation and the various attempts made to infuse new life into it through use of more modern engines after the fall of Communism. These efforts could have maintained on the market an airplane that was still appreciated, but the project backfired for the unavailability of the United Kingdom to invest money unless the new Romanian government guaranteed it.

Finally, there are a number of people to whom I would like to express my gratitude. Firstly, Professor Antonio Varsori, who believed in this project and who has always urged me to complete it, despite the delays due to the changes in my personal circumstances. I am deeply grateful not only because he encouraged me to carry on, but also because he took on the task of reading the drafts of the chapters as I finished them. Secondly, I wish to thank the Department of Political Science, Law and International Studies at the University of Padua, where I was a Postdoctoral Researcher, a position which allowed me to devote time to my studies. I am also indebted to the project "Ex post Italy. Experts and Politics on Science and Technology in Italy: A Problem of Democratic Legitimacy and International Reliability? The Interplay Between Experts and Political Decision-Makers in Italian Foreign Policy from the Crisis of the 1970s to the 'New' International and European System of the 1990s," which has contributed greatly to my work. I would also like to thank the Department of Historical Studies of the University of Milan – in particular its former director, Professor Maria Luisa Betri, and Professor Alfredo Canavero – for its contribution to publishing expenses. I am greatly indebted to Dr. Nicola Resnati, whose valuable linguistic supervision has been crucial in order to 'streamline' my impervious style.

Lastly, a special thank you and all my affection go to my friend David Burigana, who I met almost by chance many years ago, and to whom I owe my interest in aeronautics as a subject of study. Without him, this work would very probably never have seen the light of day.

Chapter 1

Western Wings for Romania

In 1968, a small booklet printed on card saw the light of day. It was well bound and the graphics were attractive and colourful. The booklet turned out to have been written by a Romanian officer, who was General Director of the Division of Aircraft Manufacturing in the Ministry of Machine Building. The volume was published by the Romanian Aviation Industry, but it was designed and printed by the British Aircraft Corporation. It contained mainly stories about Romanian aviation pioneers, but it also gave an account of an industry that, before the Second World War, had ranked sixth in Europe. The booklet ended with the British company acknowledging the fact that Romania had regained its place as a partner in the European aviation at a time when international cooperation in the field was becoming more and more important.[1]

This publication, seemingly bizarre, tangibly marked the beginning of Anglo-Romanian cooperation in the aviation industry, a complicated affair, which would outlast the historical limits of the Cold War. A concrete example of the new perspectives generated by Détente, it apparently represented a sort of oxymoron, juxtaposing the country that embraced the so-called Swinging London and one of the harshest Communist dictatorships of Eastern Europe.

And yet, the actual circumstances that the United Kingdom and Romania were going through during the 1960s, as well as the historical course that had determined them, made it possible for two antipodes to concur in a complementary relationship, supported, on the one hand, by the necessity for the British aviation industry to find new ways to guarantee its survival, on the other, by the ambitions of the Romanian regime.

---

1 Ştefan Ispaş, *Romanian Wings. A Review of Romania's Contributions in Aviation Development Through more than 60 Years*, Bucharest, IAR, n.d.

1.1 Antipodes in Détente Europe: British Industrial 'Decline' and Romanian Economic Development

The 1980s witnessed the publication of a certain number of influential essays aimed at explaining how it had been possible for the United Kingdom to find itself in such dire straits compared to other developed capitalist economies. Could it be possible that the drastic measures that Margaret Thatcher was forcing on the country found their *raison d'être* precisely in the history of the decline of the nation itself? Actually, this 'decline' that everybody quickly recognised became quite elusive and ideologically ambiguous when one looked into its causes and dynamics.

George Allen and Martin Wiener, among other historians, pinpointed a sort of cultural conservatism in the ruling classes that would prevent them from supporting national entrepreneurship as well as stopping the country from moving with the times.[2] Correlli Barnett essentially pointed to a political myopia, that is, the choice made during the post-war period to channel all the available resources into the construction of the Welfare State, undoubtedly profitable from the point of view of public approval, at the expense of restructuring the economic and productive system, a necessity that the very experience of the conflict should have made as compelling as it was evident.[3] Sidney Pollard, on the contrary, identified the reason for this decline in the inadequacy of the investments determined by stop-go dynamics,[4] citing the centrality of the weakness of the balance of payments, which had already been mentioned by the officials of HM Treasury,[5] and the government economic policies, all of which had proved counterproductive.[6]

---

2 George Allen, *The British Disease. A Short Essay on the Nature and Causes of the Nation's Lagging Wealth*, London, The Institute of Economic Affairs, 1979; Martin Wiener, *English Culture and the Decline of the Industrial Spirit, 1850-1980*, Cambridge, Cambridge University Press, 1981.

3 Correlli Barnett, *The Audit of War. The Illusion and Reality of Britain as a Great Nation*, Basingstoke, Palgrave Macmillan, 1986.

4 Typically, efforts aimed at increasing the growth rate determined an overheating of demand with ensuing overvalued exchange rate, balance of trade deficit and flight of capitals. The government was thus forced to cut public spending and/or act on fiscal policy (the 'stop'), so that actions initially intended to foster investments turned out to be counterproductive. See Barry Eichengreen, *The European Economy Since 1945. Coordinated Capitalism and Beyond*, Princeton, Princeton University Press, 2007, p. 229-232.

5 Alec Cairncross, *Managing the British Economy in the 1960s: A Treasury Perspective*, Basingstoke, Palgrave Macmillan, 1996.

6 Sidney Pollard, *The Wasting of the British Economy. British Economic Policy 1945 to the Present*, London, Croom Helm, 1982.

The historiographical debate on the decline as such is not the focus of the present work, but it is still interesting to note how these interventions, in identifying the lowest common denominator in the paralysis of the system, lay the blame heavily on the post-war Labour governments, thus portraying a certain image of Britain in the 1960s. The government led by Harold Wilson would preside over a period riddled with rhetoric and frustrated expectations, marked by a stubborn but futile defence of the pound sterling, and by the enormous waste of resources allotted, through public intervention, to the support of dysfunctional and non-competitive industrial sectors.[7] This is an image that, beyond the possible political and ideological connotations, found ample validation in historiography relating to economic policy after 1945, when it was presented as a series of missed opportunities.[8]

Starting from the 1990s, new investigations have provided a more complex and problematic picture, launching a re-evaluation of the Wilson government, designed to analyse the context of the choices made and the reasons for them.[9] Living in a period of heavy recession, historians, such as Barry Supple, have noted the apparent paradox of the contemporaneity of a century-long political debate, hinged on the idea of the incipient or ongoing decline of the country and the patency that the standard of living of the average citizen had gradually and considerably risen over the years.[10] This apparent paradox stemmed precisely from the pervasiveness of declinism, which led specialists of every discipline to diagnose an illness that the nation had contracted: the British disease,[11] to be treated perhaps

---

7 The most notable examples are the shipbuilding and the motorcar industries. See Anthony Slaven, *British Shipbuilding, 1500-2010: A History*, Lancaster, Carnegie Publishing, 2013 (especially chapter VII) and Timothy Whisler, *The British Motor Industry, 1945-1994. A Case Study in Industrial Decline*, Oxford, Oxford University Press, 1999.

8 Roger Middleton, *The British Economy Since 1945. Engaging with the Debate*, Basingstoke, Palgrave Macmillan, 2000, p. 100.

9 An early example of this tendency is Richard Coopey, Steven Fielding and Nick Tiratsoo (eds.), *The Wilson Governments 1964-1970*, London, Pinter, 1993.

10 Barry Supple, "Fear of Failing: Economic History and the Decline of Britain", in Peter Clarke and Clive Trebilcock (eds.), *Understanding Decline. Perceptions and Realties of British Economic Performance*, Cambridge, Cambridge University Press, 1997, p. 9-31. This argument is forcefully resumed and broadened by George Bernstein, *The Myth of Decline: The Rise of Britain Since 1945*, London, Pimlico, 2004.

11 Alan Booth (ed.), *British Economic Development Since 1945*, Manchester, Manchester University Press, 1995, p. 28-35.

by emulating the French economic model or by focusing attention on the apparent success of a Soviet-style planning in terms of productivity.[12]

Although Adam Smith had indeed already complained about the number of declinist papers that were proliferating in the second half of the 18th century,[13] between the 1950s and the 1960s, the greater availability of comparative statistical data, the high expectations regarding the actions of the state, which were due to the development of Welfare, along with political dialectics, produced the rhetoric of the British disease,[14] suggesting an interesting parallelism with the situation of the 1780s. In other words, insufficient investments due to bad choices related to macroeconomics,[15] the deadweight of a humanistic and amateur culture managed by an élite totally unsuitable to the demands of the modern productive world,[16] as well as the necessity for modernity and innovation for the purposes of economic growth.[17]

Even though the allegedly negative role of an elitist culture has been subject to harsh criticism,[18] here it is important to underline that the obsession with a GDP growth almost one percentage point lower than that of countries such as Germany, Japan, and Italy, in which – unlike in the United Kingdom – a relevant batch of manpower had yet to be transferred from a relatively poor agricultural sector to industry, pushed Labourists

---

12 Glen O'Hara, *Governing Post-War Britain: The Paradoxes of Progress, 1951-1973*, Basingstoke, Palgrave Macmillan, 2012.

13 Donald Winch, "A Great Deal of Ruin in a Nation", in Clarke – Trebilcock (eds.), *Understanding Decline*, *op. cit.*, p. 32-48.

14 Jim Tomlinson, "Inventing 'Decline'. The Falling Behind of the British Economy in the Post-war Years", *Economic History Review*, Vol. 49, No. 4 (1996), p. 731-757.

15 Sidney Pollard, *The Development of the British Economy, 1914-1960*, London, Arnold, 1962.

16 Charles Snow, *The Two Cultures and the Scientific Revolution*, Cambridge, Cambridge University Press, 1959. Snow was appointed parliamentary undersecretary in the Ministry of Technology during the first Wilson government. For an interesting intellectual biography of Snow, see Guy Ortolano, *The Two Cultures Controversy: Science, Literature and Cultural Politics in 1960s Britain*, Cambridge, Cambridge University Press, 2009.

17 Thomas Balogh, *Planning for Progress. A Strategy for Labour*, London, Fabian Society, 1963. Balogh, an economist, eventually became Wilson's economic counsellor. On the importance of the developments in the economic thought and the central role of professional economists, see Roger Middleton, *Charlatans or Saviours? Economists and the British Economy from Marshall to Meade*, Northampton, Edward Elgar Publisher, 1998, p. 201-281.

18 Bruce Collins and Keith Robbins (eds.), *British Culture and Economic Decline*, London, Weidenfeld and Nicolson, 1990; William Rubinstein, *Capitalism, Culture and Decline in Britain, 1750-1990*, London, Routledge, 1993.

to focus on planning, interpreted as the epitome of modernisation and efficiency.[19] The 'white heat of scientific revolution,' in the words spoken by Harold Wilson during the Labour Party Conference at Scarborough in October 1963,[20] along with a policy of selective support to industry, to be doled out according to less haphazard and more technocratic criteria, should have arrested the decline and should have claimed for the nation its rightful place among the leading economies.[21] The higher the expectations, the more searing the disillusionment, so that this very outcome resulted in new evidence for the literature on decline, as well as giving an apparent historical explanation for the principles of Thatcherism.[22]

On the contrary, more recent historiography has highlighted the coherent moments of the macro- and microeconomic interventions of the Wilson governments,[23] drawing attention to the constraints to which they were subjected[24] and to the fragmentation and the interdependence of the economic institutions in the wake of the overestimation of the real centrality (and capacity of intervention) of the state.[25] In general, research has shown how an excess of ambition should not allow us to forget that the economic performance of those years was at least satisfactory.[26] For instance, the exchange rate of the pound sterling[27] was not simply 'a fetish, a political virility symbol,'[28] unwisely defended

19 Alan Booth and Sean Glynn, *Modern Britain. An Economic and Social History*, London, Routledge, 1996, p. 241-270; Glen O'Hara, *From Dream to Disillusionment. Economic and Social Planning in 1960s Britain*, Basingstoke, Palgrave Macmillan, 2006, p. 9-36.

20 David Edgerton, "The 'White Heat' Revisited. The British Government and Technology in the 1960s", *Twentieth Century British History*, Vol. 7, No. 1 (1996), p. 53-82.

21 *Let's Go with Labour for the New Britain. The Labour Party Manifesto for the 1964 General Elections*, London, The Labour Party, 1964.

22 Glen O'Hara and Helen Parr, "The Rise and Fall of a Reputation", *Contemporary British History*, Vol. 20, No. 3 (2006), p. 295-302.

23 Glen O'Hara, "'Dynamic, Exciting, Thrilling Change': The Wilson Government's Economic Policies, 1964-70", *Contemporary British History*, Vol. 20, No. 3 (2006), p. 383-402; Hugh Pemberton, "Taxation and Labour's Modernisation Programme", *id.*, p. 423-440.

24 Jim Tomlinson, *The Labour Governments, 1964-1970. Vol. III: Economic Policy*, Manchester, Manchester University Press, 2003, p. 219-235.

25 Hugh Pemberton, "Relative Decline and British Economic Policy in the 1960s", *Historical Journal*, Vol. 47, No. 4 (2004), p. 989-1013.

26 Middleton, *The British Economy*, *op. cit.*, p. 65.

27 On the issue see Catherine Schenk, *The Decline of Sterling. Managing the Retreat of an International Currency, 1945-1992*, Cambridge, Cambridge University Press, 2010.

28 Andrew Thorpe, *A History of the British Labour Party*, Basingstoke, Palgrave Macmillan, 2008, p. 167.

by the Labour government, beyond all reasonable evidence, simply for domestic political reasons.[29] The reticence of Wilson and of the Chancellor of the Exchequer, James Callaghan, to devalue the currency was justifiable in the light of the broader international context (the further problems that the Bretton Woods system would create, the effects on those countries that still had pound sterling reserve currency, the support expected from the outside) and the conviction, later proven correct, that the problems of the British economy would not be solved over the short term.[30] It would appear that the autumn 1967 devaluation was not the inevitable result of a political mistake, but the outcome of changes in the international economy that had made it possible, up to that moment, to combine full employment and open trade – changes which led the government to settle for a middle way between deflation and expansionism.[31]

As regards the 'white heat,' David Edgerton has noted how it was, at least partially, all about an electoral strategy, destined to be superseded by new subject matter no later than the 1966 consultations, and how – practically speaking – it was not so very different from the actions undertaken by the previous Conservative governments. The 'white heat' was the conclusion, rather than the beginning, of the nationwide enthusiasm for technology. However, upon close scrutiny, and in opposition to the declinist discourse, it was apparent that the United Kingdom had not underinvested in Research and Development (R&D), but rather that it had overinvested its money and used it in the wrong areas. Thus, the 'white heat' would seem to have been an attempt to employ the experience accrued through investments in the military sector in favour of the civil sector.[32] The new Ministry of Technology (MinTech) played a crucial role in this new industrial policy, intended to reassign resources from the aerospace and nuclear sectors to other areas now considered more strategic. MinTech and its actions, generally perceived in rather negative terms, have recently been reassessed by

---

29 Alec Cairncross and Barry Eichengreen, *Sterling in Decline. The Devaluations of 1931, 1949 and 1967*, Basingstoke, Palgrave Macmillan, 1983.

30 Tim Bale, "Dynamics of a Non-Decision: the 'Failure' To Devaluate the Pound, 1964-67", *Twentieth Century British History*, Vol. 10, No. 2 (1999), p. 192-217; Scott Newton, "The Sterling Crises of 1964 and the Decision of Not To Devaluate", *Economic History Review*, Vol. 62, No. 1 (2009), p. 73-98.

31 Scott Newton, "The Sterling Devaluation of 1967, the International Economy and Post-War Social Democracy", *English Historical Review*, No. 515 (2010), p. 912-945.

32 David Edgerton, *Warfare State. Britain, 1920-1970*, Cambridge, Cambridge University Press, 2006, p. 252-265.

contemporary historians: Jim Tomlinson, for example, has pointed out that, although there were resistances and inertia, and although the overall conduct of MinTech remained all too strictly connected with the framework of 'big science,' the choice of the sectors that had to be promoted (computer technology, electronics, telecommunications, machine tools) through a selective policy of resources was reasonable and the effort altogether remarkable.[33]

As far as this topic is concerned, it has been noted that the Labour government had always harboured a certain hostility towards the aeronautical industry, which was, in a way, considered as a symbol of past bad management of R&D.[34] In an oft-quoted speech to the Society of British Aircraft Constructors (SBAC) in June 1967, Tony Benn, Minister of Technology, said that in the past 'Ministers of Aviation ran off with sums of money that made the great train robbers look like schoolboys pinching money from a blind man's tin.'[35]

In the decades after the Second World War, the aeronautical industry had undergone a progressive concentration both on the corporate level and in the processes of governmental policy decision-making.[36] The aeroplane had been the symbol of modern Britain,[37] a leading force in the scientific and technological fields, and this was supposed to translate into political influence and commercial success. Moreover, aviation represented a clear example of how, irrespective of the declinist debate, the country had invested heavily in civil technological development. The Conservative governments had spared no effort, even to the point of bickering with the Americans when meeting on the Coordinating Committee for Multilateral Export Controls (CoCom), an informal organisation endorsed by the United States at the tail end of 1949, aimed to coordinate the Allies in controlling exports to the Communist world, after the Soviet atomic

---

33 Tomlinson, *Economic Policy*, *op. cit.*, p. 94-122; see also *id.*, "After Decline?", *Contemporary British History*, Vol. 23, No. 3 (2009), p. 395-406.

34 Keith Hayward, *Government and British Civil Aerospace. A Case Study in Post-War Technology Policy*, Manchester, Manchester University Press, 1983, p. 70.

35 Anthony Benn, *Out of the Wilderness. Diaries, 1963-67*, London, Arrow, 1988, entry dated June 28, 1967.

36 See Philip Gummett, "Civil and Military Aircraft in the UK", *History and Technology*, Vol. 9 (1992), p. 203-222.

37 David Edgerton, *England and the Aeroplane. An Essay on a Militant and Technological Nation*, Basingstoke, Palgrave Macmillan, 1991, p. 83, 90-91.

explosion and the founding of the People's Republic of China,[38] precisely when it was important to sell aeronautical material.[39]

This immense effort, however, did not produce the desired results, partly because of the fickleness of governmental policy and the cancellation of development programmes,[40] in which considerable amounts of money had already been invested, and also because of an R&D approach that Edgerton did not hesitate to call 'technical leapfrogs into a commercial abyss.'[41] Thus, from the second half of the 1950s, the profitability of the aviation sector started to decrease, reaching levels considerably lower than the average for the manufacturing sector, with deleterious effects on those financial reserves that could have been invested in new projects.[42]

In December 1964, the Minister of Aviation, Roy Jenkins, announced the creation of a Committee of Inquiry into the Aircraft Industry, to be chaired by Edwin Plowden, a senior civil servant who had worked, in wartime, at the Ministry of Aircraft Production, later to become the first chairman of the United Kingdom Atomic Energy Authority. After about a year's work, Plowden's report stated that the characteristics of the aviation industry (high value added, skill-based and capital intensive) were perfectly compatible with the type of production activity considered appropriate for the United Kingdom, then in full 'white heat' climate. However, it made no sense to maintain, for reasons of prestige, an enormous sector, far too big for the narrow British market. The aviation industry had to be further rationalised, after a first round of corporate mergers had already taken place in 1959-1960,

---

38 For more on CoCom and its functioning during the Cold War, see U.S. Congress, Office of Technological Assessment, *Technology and East-West Trade*, Washington D.C., GPO, 1979, p. 151-170 and Douglas McDaniel, *United States Technology Export Control: An Assessment*, Westport, Praeger, 1993.

39 Uri Bar-Noi, "'Short List for a Long Haul': Britain's Role in the Process of Relaxing Strategic Export Controls During 1953-1954 Revisited", *Contemporary British History*, Vol. 26, No. 2 (2012), p. 173-194; Jeffrey Engel, "'We are not concerned who the buyer is': Engines Sales and Anglo-American Security at the Dawn of the Jet Age", *History and Technology*, Vol. 17 , No. 1 (2000), p. 43-67. For a reading of aviation as a technological and political struggle between the United States and Britain in the early Cold War, see Jeffrey Engel, *Cold War at 30,000 Feet. The Anglo-American Fight for Aviation Supremacy*, Cambridge (MA), Harvard University Press, 2007.

40 For example Richard Moore, "Bad Strategy and Bomber Dreams: A New View of the Blue Streak Cancellation", *Contemporary British History*, Vol. 27, No. 2 (2013), p. 145-166, and Sean Straw and John Young, "The Wilson Government and the Demise of TSR-2, October 1964 – April 1965", *Journal of Strategic Studies*, Vol. 20, No. 4 (1997), p. 18-44.

41 Edgerton, *England and the Aeroplane*, *op. cit.*, p. 94.

42 Keith Hayward, *The British Aircraft Industry*, Manchester, Manchester University Press, 1989, p. 95.

finding its *raison d'être* in the generation of profit. This would result, on the one hand, in the systematic use of international cooperation, especially in Europe, to better amortise the costs of development, to share the risk and to have a bigger commercial outlet. On the other hand, financial support from the government had to be more selective, aiming at those projects that offered actual market prospects. Finally, the report recommended that government policies be more consistent and promote exports by means of long-term loans at a reduced premium. In short, Plowden's report did not suggest abandoning the civil aviation sector, but rather focusing on what seemed more promising (medium to small aircraft, with a medium to short range), aiming to modify existing models, as opposed to developing new ones.[43]

Reactions from the industry were substantially negative, underlining, in particular, that Plowden's report sidestepped the harmful effects of the erratic government policy of defence procurement. SBAC criticised the emphasis placed on international cooperation, on the grounds that it could prove useful only when negotiated from a position of strength. The Royal Aeronautical Society maintained that the report was as thorough in its analysis of the past as it was inconclusive in forecasting the future.[44] In a conference held at the Royal United Service Institution, the Head of the Department of Aircraft Design at the College of Aeronautics in Cranfield, neatly summed up this attitude as follows:

> The popular view of the industry is that it is suffering from a serious disease. [...] I would answer by saying categorically that the aircraft industry is, second to none, the right industry for this country. First it is potentially one of our best exporting industries. [...] Then, it is a leader among science-based industries and nurtures better brains and better inventiveness to serve the whole of British industry. [...] Finally, though it is fashionable to do so, I see no reason why we should discount prestige. Great Britain has lost a lot of prestige in recent years and with the loss of prestige goes a loss of business.[45]

The practical application of Plowden's report was, however, patchy. There was, to be fair, an overall consolidation of investments in the aviation industry and a transfer of resources from the military to the civil departments sustained by the 'white heat.' The costs of R&D in the military sector dropped from 202 million pounds in 1964-1965 to 120 million in 1970-1971, while the civil sector experienced a huge

[43] *Report of the Committee of Inquiry into the Aircraft Industry*, cmd. 2853, London, HM Stationery Office, 1965.

[44] "The Industry on Plowden", *Aircraft Engineering*, Vol. 38, No. 2 (1966), p. 9.

[45] "The Future of the Industry", *Aircraft Engineering*, Vol. 39, No. 1 (1967), p. 7.

increase from 20 million to 66 million in 1968-1969.[46] It should be noted, however, that the vast majority of investments in the civil aviation industry were absorbed by Concorde which, beyond the important political implications for bilateral relations with France, and also taking into account a new application to join the European Communities by the United Kingdom, was much more similar to the reviled projects motivated by prestige rather than to a principle of cooperation animated for reasons of industrial and marketing efficiency.[47] On the contrary, in 1969, the British government decided not to take part in the Airbus consortium for political and commercial reasons, something that would have been precisely the kind of collaborative project advocated by Plowden's report.[48]

Even in terms of the rationalisation of the industry, things went differently from what was suggested in the report. In fact, although the report established favourable conditions for the acquisition by Rolls-Royce of its competitor, Bristol-Siddeley Engines, thus making it the leading producer of aircraft engines in the world in terms of number of employees and, certainly, the most advanced in Europe, the merger negotiations between the British Aircraft Corporation (BAC) and Hawker Siddeley Aviation (HSA) were not successful and the government had no precise plan in this regard. So, somewhat paradoxically, precisely in the airframe sector (in which, according to Plowden's report, rationalising was most urgent), there survived an inconvenient plurality of companies, along with their relative redundancies and the difficulty of acting as a counterweight to Rolls-Royce, especially in terms of justifying the interests of the entire aviation industry in front of the public authorities.[49]

Negotiations for the merger had failed not only because BAC's technical staff did not want to be subservient to HSA staff, but mostly

---

46 Edgerton, *Warfare State*, *op. cit.*, p. 242.

47 On the Concorde project, see Lewis Johnman and Frances Lynch, "The Road to Concorde: Franco-British Relations and the Supersonic Project", *Contemporary European History*, Vol. 11, No. 2 (2002), p. 229-251; Kenneth Owen, *Concorde and the Americans. International politics of the Supersonic Transport*, London, Airlife Publishing, 1997; David Henderson, "Two British Errors: Their Probable Size and Some Possible Reasons", *Oxford Economic Papers*, Vol. 29, No. 2 (1977), p. 159-205.

48 See David Burigana, "L'Europe, s'envolera-t-elle? Le lancement de Airbus et le sabordage d'une coopération aéronautique 'communautaire' (1965-78)", *Journal of European Integration History*, Vol. 13, No. 1 (2007), p. 91-109; Lewis Johnman and Frances Lynch, "Technological Non-Cooperation: Britain and Airbus (1965-1969)", *Journal of European Integration History*, Vol. 12, No. 1 (2006), p. 125-140; Pierre Muller, *Airbus: L'ambition européenne. Logique d'Etat, logique de marché*, Paris, L'Harmattan, 1989.

49 Hayward, *The British Aircraft Industry*, *op. cit.*, p. 121-125.

because the former company had witnessed a significant improvement in its financial situation through a massive 65-million-pound military contract with Saudi Arabia and to the success of its short-haul jet airliner, the BAC 1-11.[50] Developed on the basis of an initial order of ten aircraft from British United Airways, the plane had some features that made it very appealing: it was relatively cheap, with constructive solutions that ensured a high resistance to fatigue and easier access for inspections. It was also suitable for short-haul flights and had a high utilisation factor. It was, on the whole, a machine that, under conditions of economic efficiency, could effectively serve routes which, until then, had used the less comfortable and slower turboprop planes, thus facilitating in particular the booming industry of package holidays, with flights included, carrying passengers from North-western Europe to seaside resorts. The success of the product was demonstrated by the fact that, unlike other British aircraft, the BAC 1-11 was sold mainly on the export market.[51]

The characteristics of the BAC 1-11 were one of the reasons for a special relationship between Britain and Romania in the aviation industry that would last more than twenty years, embracing the whole parabola of Détente and the new phase of bipolar tension in the 1980s, terminating a few years after the end of the Communist dictatorship. The longevity of this relationship is all the more remarkable when one considers that, in its beginnings, during the 1960s, the two countries appeared, in many ways, to be poles apart in Détente Europe.

Romania, in fact, after the arrival of the Red Army and after the seizure of power by a Communist Party divided into factions, had long remained one of the satellites most closely aligned to Moscow. The General Secretary of the Romanian Workers' Party (Partidul Muncitoresc Român, PMR), Gheorghe Gheorghiu-Dej, was one of the most violent opposers of Titoism after the expulsion of Yugoslavia from the Cominform.[52] Romania had the opportunity to show its loyalty to the USSR during the events

---

50 Charles Gardner, *British Aircraft Corporation*, London, Batsford, 1981, p. 149-160, 244-251.

51 Norman Barfield, "The BAC One-Eleven Family", *Aircraft Engineering*, Vol. 42, No. 3 (1970), p. 6-14.

52 See his speech, "Le Partie communiste yougoslave au pouvoir des assassins et des espions", at the third Cominform Conference in 1949, in Giuliano Procacci, *The Cominform. Minutes of the Three Conferences (1947-1949)*, Milan, Feltrinelli, 1994, p. 971-976.

in Hungary in 1956, in part motivated by the concern of the potential backlash on the Hungarian minority in Transylvania.[53]

This scenario was destined to change gradually and in a discontinuous way, turning Romania into a sort of maverick of the Communist world and its leaders into the supposed champions of autonomy from Moscow, finding in the long Détente period the opportunity to carry out an attempt to distance itself from the Soviet Union and to re-approach the Western world.[54] According to Vladimir Tismăneanu, the ultimate reason for this process was the fear of the Romanian leadership regarding the effects of a Krushchevean de-Stalinisation and the retention of power by the PMR, especially given the fact that the period during which Romania tried to get rid of the Soviet stranglehold coincided with that of anti-Stalinist campaigns all over the bloc. Breaking ranks internationally, therefore, was crucial in order to maintain a sort of Stalinist compliance within Romania, thus building an effective platform played out on the concepts of sovereignty, independence and industrialisation as a matter of dignity and national pride. It was a platform that, while careful not to overstep the mark with Moscow, guaranteed an effective measure of internal cohesion to the PMR.[55]

There are various assessments of the actual degree of Romanian independence from the USSR and, most recently, innovative interpretations that attempt to go further, reading Romanian politics as an effort of more general 'anti-hegemonism,' that is to say, the refusal to attribute to two superpowers a special responsibility in international affairs, determined not so much by the perception of an external opportunity to be exploited, but by the desire to avoid new forms of dependence.[56] That being said, starting from the abandonment of the Romanian territory by the Soviet troops,[57] it certainly became more plausible to follow a policy that would

---

53 Dennis Deletant, "Romania, 1948-1989. An Historical Overview", in *Romania and the Warsaw Pact: 1955-1989*, Cold War History International Project, Working Paper No. 43 (2004), p. 2-12.

54 Vladimir Tismăneanu, *Gheorghiu-Dej and the Romanian Workers' Party: From the De-Sovietisation to the Emergence of National Communism*, Cold War History International Project, Working Paper No. 37 (2002); Mioara Anton, *Ieşirea din cerc. Politica externă a regimului Gheorghiu-Dej*, Bucharest, INST, 2007; Constantin Moraru, *Politica externă a României, 1958-1964*, Bucharest, Ed. Enciclopedică, 2009.

55 Vladimir Tismăneanu, *Stalinism for All Seasons. A Political History of Romanian Communism*, Berkeley, University of California Press, 2003, p. 168-180.

56 Elena Dragomir, "The Perceived Threat of Hegemonism in Romania During the Second Détente", *Cold War History*, Vol. 12, No. 1 (2012), p. 111-134.

57 See Sergiu Verona, *Military Occupation and Diplomacy. Soviet Troops in Romania: 1944-1958*, London, Duke University Press, 1992.

give preference to independence and national interests, as was shown by the determination to minimise the powers of the Warsaw Pact agencies, to the point of claiming the right of Bucharest to decide whether or not to intervene in case of aggression from the West and, if so, on what terms, while also seriously considering the possibility of exiting the organisation *à la française*.[58]

Even prior to the dynamics of the Warsaw Pact, the maturation of the Romanian national-Communism can be identified in those of economic development. The Soviet-style planned systems, resembling wartime economies, allowed for the concentration of resources and production factors by administrative means, thus ensuring the rapid industrialisation of more or less underdeveloped territories, with average growth rates among the highest ever recorded. An intrinsic component of this model was the isolation of internal activities from external influences, through the state monopoly on foreign trade – something that Joseph Stalin turned into the ideological principle of separating the socialist and capitalist markets.[59] Exchanges within the bloc took place as part of the bilateral relations between the USSR and the individual satellites under the umbrella of the Council for Mutual Economic Assistance (CMEA), born thanks to a Romanian initiative, originally intended to provide the country with the necessary assistance for industrialisation.[60]

That of the CMEA was a cooperation based on the principle that every member had the right to express their interest in a given proposal and that this could not be forced upon them without their consent. Resembling more the Organisation for European Economic Cooperation than the Common Market, the CMEA basically remained in a vegetative state until the death of Stalin, while each of the bloc countries was pursuing the

---

58 Vojtech Mastny and Malcolm Byrne (eds.), *A Cardboard Castle? An Inside History of the Warsaw Pact, 1955-1991*, Budapest, Central European University Press, p. 210-211, 264-269. According to Garthoff, in October 1963, Foreign Minister Corneliu Mănescu told Dean Rusk that, since Romania had not been consulted on the Soviet decision to deploy missiles at Cuba, his leaders would consider themselves neutral in a conflict generated by any such unilateral action. See Raymond Garthoff, "When and Why Romania Distanced Itself from the Warsaw Pact", *CWIHP Bulletin*, No. 5 (1995), p. 111. For a more general overview, see Mioara Anton, *Introduction*, in *România și Tratatul de la Varșovia. Conferințele miniștrilor Afacerilor Externe și ale adjuncților lor (1966-1991)*, Bucharest, Alpha, 2009, p. XLVI-LI.

59 Iosif Vissarionovich (Džugašvili) Stalin, *Economic Problems of Socialism in the USSR*, Moscow, Foreign Languages Publishing House, 1952.

60 Elena Dragomir, "The Creation of the Council for Mutual Economic Assistance as Seen from the Romanian Archives", *Historical Research*, 2014, DOI: 10.1111/1468-2281.12083.

same goals of economic development, endlessly reverting to a system that, behind the sensational growth rates, reproduced the backwardness and obsolescence of a production system tailored to the early 20th century.[61]

After Stalin's death, the intrinsic limits a of purely extensive industrialisation and the changing political climate led to interventions, both at the CMEA level, stressing a strong need for the coordination of national planning that could have solved the problem of the overlapping of productive specialisations, and at national level, with the opening of a season of economic reforms. In both cases, Romania's route was peculiar. Certainly, the Bucharest establishment took advantage of the changed climate to liquidate the SovRom experience, joint ventures which the Soviets basically used to exercise a pseudo-colonial control over the most important sectors of the Romanian economy.[62] The government also fully participated in the rediscovery of international trade as a stimulating factor for technological progress and, therefore, for increased productivity.[63]

Having said that, for Gheorghiu-Dej and his successor the distinction between the USSR and the Soviet model remained a valid one, that is to say, that Romania's relative non-alignment with Moscow in foreign policy did not mean the cessation of a commitment in favour of a Leninist-Stalinist industrialisation.[64] The top priority thus remained the development of the metallurgical, chemical and mechanical industries, to the point where Romania came to outdo the United States and the Soviet Union in the per capita production of steel at a time when energy costs, concerns for the environment and the availability of new materials led to a drop in global steel consumption.[65] In the light of this commitment, which simultaneously became an economic objective and a reason for internal legitimacy for the PMR, the attempts to revamp the CMEA as a supranational organisation, along with those to ensure product specialisation within the bloc, were read as a threat to force Romania to remain an agricultural country for the benefit of its sister countries. So,

---

61 Ivan Berend, *An Economic History of Twentieth-Century Europe. Economic Regimes from Laissez-Faire to Globalization*, Cambridge (MA), Cambridge University Press, 2006, p. 166-175; Robert Bideleux and Richard Taylor (eds.), *European Integration and Disintegration: East and West*, London, Routledge, 1996, p. 174-204.

62 Florian Banu, *Asalt asupra economiei României, de la Solagra la SOVROM (1936-1956)*, Bucharest, Nemira, 2004.

63 Jozef Van Brabant, *Socialist Economic Integration*, Cambridge, Cambridge University Press, 1980, p. 156-159.

64 Deletant, *Romania, 1948-1989*, *op. cit.*, p. 16-17.

65 Bogdan Murgescu, *România şi Europa. Acumularea decalajelor economice (1500-2010)*, Iaşi, Polirom, 2010, p. 343-346.

starting from the end of the 1950s, the Romanian government would increasingly, and more and more patently, oppose an evolution considered detrimental to its political and economic independence, *de facto* managing to successfully boycott it.[66]

Furthermore, on May 18, 1959, the PMR Politburo decided to start obtaining technology that the USSR had not previously wanted or been able to offer to Romania from the capitalist countries of Western Europe. It was believed, in fact, that the economic situation in the West and competition among potential suppliers to enter new markets would make it possible to limit the amount of investments and to obtain particularly favourable financial conditions, especially the possibility to pay for imports of technology thanks to the export of goods produced by applying these very same new technologies. In theory, therefore, it was supposed to be an operation with a minimal outlay of hard currency – albeit with a series of objective constraints – and bound to be fully paid off, offering Romania plant and production processes that would enable the country to set the bar for quality, an aim towards which, of course, workers, technicians and engineers had to be mobilised.[67]

The emergence of a 'national policy'[68] was therefore born out of the blending of an economic development with very specific connotations, the internal political consolidation of the regime and the affirmation of

---

66 Lavinia Betea, *Alexandru Bârlădeanu despre Dej, Ceaușescu și Iliescu. Convorbiri*, Bucharest, Ed. Evenimentul Românesc, 1997, p. 147; Elena Dragomir, "Romania's Participation in the Agricultural Conference in Moscow, 2-3 February 1960", *Cold War History*, Vol. 13, No. 3 (2013), p. 331-351; Maria Mureșan, "Romania's Integration in COMECON. The Analysis of a Failure", *The Romanian Economic Journal*, Vol. 9, No. 30 (2008), p. 27-58; Liviu Țăranu, *România in Consiliul de Ajutor Economic Reciproc, 1949-1965*, Bucharest, Ed. Enciclopedică, 2007. On the evolution and eventual failure of 'socialist integration,' see Randall Stone, *Satellites and Commissars. Strategy and Conflict in the Politics of the Soviet-Block Trade*, Princeton, Princeton University Press, 1996, p. 49-71; Simon Godard, « Construire le bloc de l'Est par l'économie? La délicate émergence d'une solidarité internationale socialiste au sein du Conseil d'aide économique mutuelle », *Vingtième Siècle*, No. 109 (2011), p. 45-58; André Steiner, "The Council of Mutual Economic Assistance. An Example of Failed Economic Integration?", *Geschichte und Gesellschaft*, Vol. 39, No. 2 (2013), p. 240-258.

67 The National Archives of Romania (ANR), Central Committee of the Romanian Communist Party (CC), Chancellery, 1959_19: nota privind angajarea – din țările capitaliste – a unor instalații complexe, 06.05.59, ff. 49-60; protocol No. 10 al ședinței Biroului Politic din 18 mai 1959, ff. 1-17.

68 Paul Niculescu-Mizil, *O linie politică naționala*, in *id.*, *O istorie trăită*, Vol. 1, Bucharest, Ed. Enciclopedică, 1997, p. 9-38.

Romania in terms of international relations.[69] It was, in short, a dynamic of mutual reinforcement between these three factors. If the mediocre consumer goods coming from the CMEA were considered more than sufficient, the Romanian leadership wanted to adopt high-quality, modern production equipment, which was available only in the West, to ensure the industrial future of the country, rightly believing that it would be offered favourable credit conditions. Thus, the share of imports of machinery coming from capitalist countries soared from 9.5% in 1958 to 25.7% in 1960, reaching an average of 47.3% between 1966 and 1973[70] and, in general, the rapid development of foreign trade – in which the mechanical construction sector represented the most dynamic element – played a crucial role in supporting high rates of economic growth starting from the mid-1970s.[71]

The growth in trade, in turn, passed through the revival of diplomatic relations with the Western countries and through the resistance to the projects of economic integration promoted by the Soviets within the CMEA; quite naturally, Western forthcomingness and benevolence were directly proportionate to the degree of perceived autonomy from Moscow and they generated a measure of economic support that Maoist China, decidedly on a collision course with the Soviet Union, was objectively unable to give to Romania. In this game of action and reaction, Gheorghiu-Dej embraced the idea of Palmiro Togliatti's polycentrism[72] and the concept of equal dignity for all Communist parties put forward by China, managing to introduce the PMR as the champion of Romanian national interests.[73] This process culminated, following a fierce attack on the so-called 'Valev Plan' (a Soviet project of cross-border integration of the Lower Danube basin),[74] in the declaration of the Central Committee of the

69 Dan Cătănuş, „Divergenţele româno-sovietice din CAER şi consecinţele lor asupra politicii externe a României", I, *Arhivele Totalitarismului*, No. 1-2 (2005), p. 68-80.

70 John Montias, "Romania's Foreign Trade Between East and West", in Paul Marer and John Montias (eds.), *East European Integration and East-West Trade*, Bloomington, Indiana University Press, 1980, p. 326-329.

71 David Turnock, *The Romanian Economy in the Twentieth Century*, London, Croom Helm, 1986, p. 231-241; Nicolae Sută (ed.), *Istoria comerţului exterior românesc. O prezentare sintetică*, Bucharest, Eficient, 1998, p. 165-272.

72 Palmiro Togliatti, *Il memoriale di Yalta*, Palermo, Sellerio, 1988.

73 Tismăneanu, *Stalinism for All Seasons*, *op. cit.*, p. 178-179; Dennis Deletant, "'Taunting the Bear': Romania and the Warsaw Pact, 1963-89", *Cold War History*, Vol. 7, No. 4 (2007), p. 496.

74 Emil Valev, "Problems of Economic Development of the Danube Region of Romania, Bulgaria and the Soviet Union" (in Russian), *Вестника Московского университета*, No. 2 (1964), p. 56.

PMR in April 1964. This stated some basic principles of Romanian policy (including enduring respect for independence and national sovereignty, equality among States, non-interference in internal affairs, cooperation based on mutual benefit) and rejected any hierarchical distinction between Communist parties, each of which was asked to identify the best path to the construction of socialism in their respective countries.[75]

This declaration of independence epitomised the Communist leadership's reactivation of the nationalist debate. The reactivation was implemented through the initiative of the State, which leveraged on the Russophobic anti-Sovietism of the Romanians and played on the reappearance of patriotism in the humanities (the Latin origin of the Romanian language, the Daco-Romanian continuity, etc.)[76] Within this debate, industrial development became a yardstick of national dignity, presenting a series of elements that partially reached back to the interwar period, for example with the principle '*Prin us înşine*' (meaning, by ourselves alone) of the national-liberal economic doctrine applied to the appropriation of technology and of production processes.[77] This would help to explain the relatively broad support initially received by the economic options of the regime.[78]

All this, of course, inevitably had consequences in terms of increased animosity with the sister countries. During the Romanian-Bulgarian-Soviet joint naval manoeuvres of February-March 1965, an anonymous Soviet officer burst out: "[…] we have delivered you from the Turks, we have delivered you from the Germans, we built your industry and how come you are never grateful?"[79] Relations with the capitalist countries, then, was a particularly delicate and thorny topic, as one can tell from Gheorghiu-Dej's tirade on the occasion of a meeting with Tito, the very

---

75 Florian Banu and Liviu Ţăranu, *Aprilie 1964 „Primăvara de la Bucureşti". Cum s-a adoptat "Declaraţia de independenţă" a României?*, Bucharest, Ed. Enciclopedică, 2004, p. VII-L; Paul Niculescu-Mizil, *Împotriva dezmembrării economice şi teritoriale a României*, in *id.*, *O istorie trăită*, Vol. 1, *op. cit.*, p. 235-250.

76 Irina Gridan, « Du communisme national au national-communisme: réactions a la soviétisation dans la Roumanie des années 1960 », *Vingtième Siècle*, No. 109 (2011), p. 113-127.

77 Robert Bideleux and Ian Jaffries, *A History of Eastern Europe: Crisis and Change*, London, Routledge, 1998, p. 551-552.

78 John Montias, *Economic Development in Communist Romania*, Cambridge (MA), MIT Press, 1967.

79 ANR, CC, Chancellery, 1965_194: Report from the Minister of the Armed Forces to the CC, 01.04.65, f. 21.

same man who in the past he had accused of being a spy on the payroll of the imperialists:

> Comrade Gheorghiu-Dej has also highlighted the necessity to expunge distrust, in the relations between the member parties. One cannot understand how anyone can say that the leaders of our party would 'sell their soul to the devil,' if we were to develop economic and cultural relations with capitalist countries. Our party has fought in illegality and under difficult conditions, it has supported numerous sacrifices, its leaders have suffered police terror perpetrated by the bourgeois-aristocratic regime, but they have not sold their souls. […] How can we now say that 'we sell our souls to the devil,' that we will let ourselves be influenced by the capitalists, that we will adopt their system, etc.?[80]

This time Gheorghiu-Dej was sincere. If it is possible to find a common denominator for the reform efforts of the planned economies in the 1960s, it mainly consisted in a sort of decentralisation of the decision-making processes and in the re-introduction of some market mechanisms. From this point of view, Romania represented an exception to the rule, in the sense that the changes were rare, even insignificant, intended more to improve the efficiency of the traditional planning system rather than to modify it.[81] In other words, according to János Kornai's model, Romania remained faithful to the 'classical socialism' of Stalinist origin and it had no sympathy for the 'reform socialism.'[82]

Nicolae Ceaușescu inherited and continued the policy set out by Gheorghiu-Dej, adding to it his own peculiar personality traits. The backwardness of the country – and, conversely, industrialisation – became an obsessive concern, inflected with a tendency to gigantism frequently disconnected from pragmatic considerations. The ambition and vanity of the *Conducător* (meaning leader) urged him to aim at transforming Romania into a major industrial power in the shortest possible time, through a massive mobilisation of society and the infusion of all the production

---

80 ANR, CC, Foreign Relations, 1964_49: convorbirea tov. Gheorghiu-Dej cu tov. Tito, 07.09.64, ff. 15-16.

81 Derek Aldcroft and Steven Morewood, *Economic Change in Eastern Europe since 1918*, Aldershot, Elgar, 1995, p. 113-115; Michael Kaser and Iancu Spigler, "Economic Reform in Romania", in Alec Nove, Hans-Hermann Hohmann and Gertrud Seidenstecher (eds.), *The East European Economies in the 1970s*, London, Butterworths, 1982, p. 253-279; Iancu Spigler, *Economic Reform in Romanian Industry*, Oxford, Oxford University Press, 1973.

82 Janos Kornai, *The Socialist System. The Political Economy of Communism*, Oxford, Clarendon, 1992.

factors considered necessary for the purpose. In this sense, foreign policy – beyond being a factor of internal legitimacy and personal recognition – was also an integral part of an economic programme that needed technology and Western credits more than ever.[83] Megalomania, centralism and the interference, even in small details, of the 'Communist Pharaoh'[84] and his wife Elena would play a primary role in Romania's economic disaster, as well as in its final political isolation, allowing, after 1989, the *nomenklatura* survivors, who had been the protagonists of the 'new-old' Romanian political course, to oppose the sensible and solid economic policy of the Gheorghiu-Dej times to the disorderly and unreasonable one supported by Ceauşescu, minimising their elements of continuity.[85]

During the July 1965 Congress, when Ceauşescu was appointed Secretary-General and the PMR was once again called the Communist Party (Partidul Comunist Român, PCR), so that the 4th Congress of PMR became the 9th of the PCR, the new five-year plan (1966-70) was launched. The ambitious goals of industrial development set by the plan required formidable imports of raw materials, plant and machinery.[86] By the end of 1967, the possibility that this economic policy could lead to the implementation of incisive reforms had been set aside. On the contrary, the Central Committee explicitly ruled out a shift from direct target planning to a parametric system, that is to say, more indirect and decentralised. So, all the interventions, even if significant in terms of openness to international economic cooperation, such as the directives given by Ceauşescu at the 1969 PCR Congress,[87] or the 1971 Foreign Trade Law,[88] which opened the possibility for Western companies to create joint ventures in Romania, did not change the essence of the planning mode – much less the nature of the regime.[89]

---

83 Trond Gilberg, *Nationalism and Communism in Romania. The Rise and Fall of Ceauşescu's Personal Dictatorship*, Westview, Boulder, 1990, p. 61-64, 72-115, 210-211.

84 Tismăneanu, *Stalinism for All Seasons*, *op. cit.*, p. 189.

85 Betea, *Alexandru Bârlădeanu*, *op. cit.*, p. 152-153; *Id.*, *Maurer şi lumea de ieri. Mărturii despre stalinizarea României*, Arad, Ed. Felix, 1995, p. 226-227.

86 *Directivele Congresului al IX-lea al PRC cu privire la dezvoltarea economiei naţionale în perioda 1966-1970*, Bucharest, Ed. Politică, 1965.

87 *Extinderea relaţiilor economice, cooperarea în producţie – o necesitate a progresului material, un factor al pacii şi colaborarii internaţionale*, in *Congresul al X-lea al Partidului Comunist Român, 6-10 august 1969*, Bucharest, Ed. Politică, 1969, p. 42-44.

88 Lege 1/1971 cu privire la activitatea de comerţ exterior, de cooperare economică şi tehnico-ştiinţifică a Republicii Socialiste România, available at http://lege5.ro (last accessed on July 29, 2015).

89 Mary Fischer, *Nicolae Ceauşescu. A Study in Political Leadership*, Boulder, Rienner, 1989, p. 109-113; Wlodzimierz Brus, "1966 to 1975: Normalisation and Conflict",

Although the official Romanian anti-hegemonism, at this stage, was more than enough in the eyes of Western countries to overshadow the nature of a Stalinism that came across as particularly brutal and stifling, and although relations with the West had a crucial economic and political functionality, it should still be noted that the United Kingdom did not feature among Romania's key partners. The Romanian Foreign Minister at the time, Corneliu Mănescu, recalls how the two simultaneous and most important trajectories of foreign policy were, on the one hand, approaching China and, on the other hand, an openness towards Western countries, with France and the Federal Republic of Germany (FRG) in the foreground.[90] This perspective has been substantially retained by recent historiography, which has considered the integration of the CMEA and the Soviet demand to put up a common ideological front against Maoist China as the immediate reasons for the redefinition of Romanian foreign policy.[91]

So attention was paid to Franco-Romanian relations after the Second World War,[92] especially to Prime Minister Ion Gheorghe Maurer's visit to Paris in 1964, which resulted in a change of pace in bilateral relations between the two countries[93] with the signing of an agreement for long-term cooperation that would grant Romania access to technology and industrial installations on particularly favourable terms.[94] Important contributions have been dedicated to relations with the FRG – with which Romania was the first country behind the Iron Curtain to establish diplomatic relations in 1967 – and Italy, respectively its second and third largest trading partners after the Soviet Union.[95] In contrast, the United Kingdom would never have an equally decisive role, both because its economic

---

in Michael Kaser (ed.), *The Economic History of Eastern Europe 1919-1975, Vol. III: Institutional Change within a Planned Economy*, Oxford, Clarendon, 1986, p. 222-228.

90 Corneliu Mănescu, *Convorbiri nedeterminate. Corneliu Mănescu în dialog cu Lavinia Betea*, Bucharest, Polirom, 2001, p. 159-168.

91 Elena Dragomir, *Cold War Perceptions: Romania's Policy Change Towards the Soviet Union, 1960-1964*, Newcastle upon Tyne, Cambridge Scholars Publishing, 2015.

92 Gavin Bowd, *La France et la Roumanie communiste*, Paris, L'Harmattan, 2008.

93 Irina Gridan, « Parallélismes et convergences (III). Ion Gheorghe Maurer a Paris a l'été 1964: Realpolitik et offensive du sourire », *Arhivele Totalitarismului*, Vol. XVII, No. 64-65 (2009), p. 99-122.

94 ANR, CC, Foreign Relations, 1964_78: notă privind discuțiile cu partea franceză in vederea încheierii unui acord comercial de durată, 09.11.64, ff. 1-8.

95 Dumitru Preda, Claudiu Florian and Ottmar Trasca (eds.), *România-Republica Federală Germană, (1966-1967). Vol. 1: Începutul relațiilor diplomatice*, Bucharest, Ed. Enciclopedică, 2009; Giuliano Caroli, *La Romania nella politica estera italiana, 1919-1965. Luci e ombre di un'amicizia storica*, Milan, Negard, 2009, p. 429-483.

system would never be able to offer better products and services than its competitors, which were geographically closer to Romania, and also because of the greater ideological distance, especially with Conservative governments.[96]

As a matter of fact, in the United Kingdom very little was said about East-West relations in the 1964 elections and the attention of the politicians of the day, and later of historians, would be engaged in other foreign policy issues, such as the entrance into the European Communities, the issue of nuclear sharing in NATO, the independence of Rhodesia and the risk of a civil war in Nigeria, the withdrawal from East of Suez and the Vietnam War.[97] Obviously, for reasons both historical and circumstantial, any British government would be in favour of the development of commercial relations with the East: the views of Richard Cobden and the treaty with France of 1860 – according to which trading strengthens a mutual interest in prosperity, while also decreasing the risk of war between two countries that are close because of a mutual supplier-customer relationship – were, in fact, made more urgent by those declinist anxieties and by stiff competition from continental European countries that were expanding at an ever-increasing rate.[98]

By the mid-1960s, the prevailing view in Whitehall was that commercial, cultural and scientific contacts served to support the interrelated processes of Détente and liberalisation within the bloc, which were, in short, functional to an 'evolution' meant as a progressive erosion of Communism and a dissolution of the Cold War. Wilson himself had often talked about the importance of trading with the East to support economic growth in the United Kingdom, but tended to focus primarily on the Soviet Union,

---

96 Gilberg, *Nationalism and Communism in Romania, op. cit.*, p. 221.

97 Rhiannon Vickers, *Labour's Foreign Policy*, Manchester, Manchester University Press, 2011; Kristan Stoddart, *Losing an Empire and Finding a Role. Britain, the USA, NATO and Nuclear Weapons, 1964-70*, Basingstoke, Palgrave Macmillan, 2012; Helen Parr, *Britain's Policy Towards the European Community. Harold Wilson and Britain's World Role, 1964-1967*, London, Routledge, 2006; Philip Alexander, "The Tale of the Two Smiths: the Transformation of Commonwealth Policy, 1964-70", *Contemporary British History*, Vol. 20, No. 3 (2006), p. 303-321; Sylvia Ellis, *Britain, America, and the Vietnam War*, Westport, Praeger, 2004; Saki Dockrill, *Britain's Retreat from East of Suez. The Choice Between Europe and the World?*, Basingstoke, Palgrave Macmillan, 2002.

98 Brian White, *Britain, Détente and Changing East-West Relations*, London, Routledge, 1992, p. 120-129, 162-169; Martin Ceadel, "Cobden and Peace", in Anthony Howe and Simon Morgan (eds.), *Rethinking Nineteenth-Century Liberalism: Richard Cobden Bicentenary Essays*, Aldershot, Ashgate, 2006, p. 189-207.

showing little interest in its satellites.[99] This was partly because of his past experiences, and because he was obeying a certain conception of the role of his country between the two superpowers. Moreover, Anglo-Romanian relations were actually at a low ebb. Poisoned by disputes resulting from the expropriations of 1948, which damaged British interests in the oil sector, they resumed only in 1959, as part of the opening of Romania to the West, when Bucharest paid partial compensation in return for a trade agreement in 1960. It was a dynamic that proved to be very similar to what had happened with other bloc countries starting from the mid-1950s, albeit comparatively late and lukewarm.[100] There had been contacts with Dunlop and Imperial Chemical Industries for the creation of a tyre factory in Bucharest and one for the production of polyethylene in Ploieşti,[101] but paradoxically the exchanges had grown thanks to the sharp increase in British imports from Romania, causing the visible trade balance to start recording, from 1964 onwards, a deficit of half a million pounds destined to rise to 16 million in 1967.[102]

Yet, by the end of the decade, Benn found himself in Bucharest to discuss with the First Deputy of the Romanian Council, Ilie Verdeț, how, through technological cooperation, the interests of the two countries would be mutually benefited:

> Romania's national interest was in building up her industrial strength as quickly as possible though access to the research and advanced technology of more advanced nations thus guaranteeing her a future in the modern industrial world and reinforcing her political independence. The Minister said that [the British national objectives] were to improve political relations with Romania; to go beyond active coexistence to vigorous cooperation in science and technology. […] The development of a partnership in technology was therefore not just a matter of trade in the short term; market factors decided whether or not a technology could be developed at all. We were therefore trying to see that our technological agreements gave us a secure forward market.[103]

---

99 Geraint Hughes, *Harold Wilson's Cold War. The Labour Government and East-West Politics*, 1964-1970, Woodbridge, Boydell and Brewer, 2009, p. 41-57.

100 Malcolm Hill, *East-West Trade, Industrial Co-operation and Technological Transfer. The British Experience*, Aldershot, Gower, 1983, p. 7-10.

101 Gheorghe Gaston Marin, *În serviciul României lui Gheorghiu-Dej. Însemnări din viață*, Bucharest, Evenimentul Românesc, 2000, p. 209.

102 The National Archives of the United Kingdom (UKNA), FCO 28/250: note by Mewes, 29.01.68.

103 UKNA, FCO 28/257: note of a meeting between Benn and Verdeț held on 06.06.68.

The British aviation industry and the plans for modernisation engineered by the Romanian airline TAROM were the agents of this sea change and marked the beginning of a complex and controversial relationship that would bind the two countries until after the fall of Communism.

1.2 Competition for TAROM's Modernisation: the Choice in Favour of BAC

A corollary of Romania's opening towards the West was the realisation that tourism was a profitable venture. This was a way to endorse the resources of the country and, above all, a valuable channel to obtain hard currency, something that was more important than ever for the industrialisation effort. At the 9th Congress of the PCR there was talk of the opportunity to develop tourism through the upgrading of receptive facilities and a significant growth – both quantitative and qualitative – of air transport, to be carried out by acquiring modern equipment and by improving the airport network, starting with the construction of the new Bucharest Otopeni airport.[104] A national tourist office was created to offer organised tours in association with Western travel agencies, attractions were advertised and national information points abroad were opened. Between 1965 and 1969 the annual growth rate of foreign tourists (including those from sister countries) was an average of 107%.[105] At the 10th Congress, in 1969, Maurer said:

> The great extension that tourism is going through today in the world has turned it, as you very well know, into an important means to increase the exchange reserves. As Comrade Ceaușescu has shown, the landscape and climate of our country, the number of health resorts at the sea and in the mountains, the valuable monuments at its disposal and the hospitality of the Romanian people are all major factors in the intensification, in the years to come, of international tourism in Romania.[106]

Tourists, however, especially those coming from the West, had to be brought to Romania quickly and comfortably. If for top Romanian

---

[104] *Congresul al IX-lea al Partidului Comunist Român, 19-24 iulie 1965*, Bucharest, Ed. Politică, 1965, p. 773, 785.

[105] Murgescu, *România și Europa, op. cit.*, p. 376; David Turnock, "Romania", in Derek Hall, *Tourism and Economic Development in Eastern Europe and the Soviet Union*, London, Belhaven Press, 1991, p. 203-219.

[106] *Congresul al X-lea al PCR, op. cit.*, p. 112.

executives it was still completely normal to go to Paris or London by train, thus wasting a couple of days to cross Europe, this would not work for those who had bought a package holiday and wanted to spend a couple of weeks on the Black Sea. What was needed was planes that were safe from the technical point of view and that complied with international standards of comfort. In the period between the two wars, on the initiative of the State, Romania had created a relatively large aviation industry to meet the demands of the armed forces. By acquiring licences from France, Poland, the United States, Italy and Germany, the Industria Aeronautică Română (IAR) in Braşov had been able to produce military aircraft that garnered a certain technical renown, although engines and some of the components had to be imported, but little or nothing of it survived the Second World War.[107]

The Romanian state had not dedicated much attention to the civil aviation sector and, from 1946, it was absorbed into one of the SovRoms, the TARS (Societatea de Transporturi Aeriene Româno-Sovietice), which operated a small fleet of Lisunov Li-2 (the Soviet version of the American DC-3 that had entered production in the 1930s) on a limited network of domestic routes and which also flew to the capitals of some of the sister countries bordering Romania.[108] In 1954, the SovRoms were dissolved and TAROM (Societatea de Transporturi Aeriene Române) was formed. On January 25, 1955 an intergovernmental agreement with the USSR was signed; it called for the inauguration of a regular connection between Bucharest and Moscow and for the training of Romanian flight personnel by the Soviets.[109] Over the next decade, TAROM built up a fleet of more modern four-engine turboprop Ilyushin Il-18s and, in addition to the capitals of Eastern Europe, it started serving the routes connecting Bucharest to Zurich, Paris, Vienna, Copenhagen, Frankfurt, Athens, Brussels and London.[110]

In the mid-1960s, therefore, the next logical step was to equip TAROM with faster and more comfortable jet aircraft and, consistently with the approach that Romania had adopted in those years, talks were held not only with the Soviet Union, but also with the major Western aircraft manufacturers, particularly those based in France and the United Kingdom. Already in December 1964, a delegation of HSA travelled to

---

107 Ion Dumitraşcu, *Industria Aeronautică Română Braşov (1925-1945)*, Bucharest, Domino, 2008.

108 "The World Airlines", *Flight International*, 06.03.53, p. 313.

109 "Iron Curtain Airlines News", *Flight International*, 11.02.55, p. 184-185.

110 "World Airlines Survey", *Flight International*, 14.04.66, p. 637.

Bucharest to offer two Tridents to be used for charter services, but the favourites to secure the Romanian order seemed to be the French of Sud Aviation with their Caravelle. Only a year later, in December 1965, did BAC present an offer for four 1-11s – planes that it was believed the Romanians knew little about from a technical standpoint.[111]

It should also be stressed that the interest in Western aeroplanes was not just the desire to spite the Soviets; indeed, the possibility of buying the new Tupolev Tu-134 was being seriously considered – if only because of the financial benefits of an investment in roubles that would then have generated profits in Western currencies. The problem was that the Soviets, since 1965, had not been able to offer the Romanians precise delivery dates for the supply of aircraft, to the point where it was necessary for them to unwillingly buy four more Il-18s to cover at least partially the upgrading of the fleet specified in the five-year plan of 1966-70. Furthermore, there were technical problems. Undoubtedly, the Il-18 had the advantage of a high lift/drag ratio and an undercarriage designed to land on relatively bumpy runways, but the attention devoted to comfort limited the number of seats available to well below the Romanian specifications and, above all, the machine was not equipped with thrust reversers. This last aspect put the plane below Western standards and, in the event of ice or frost, the safety of passengers could have been jeopardised.[112]

That same December 1965, Prime Minister Maurer astonished the British Ambassador Leslie Glass by summoning him not for Christmas greetings, but to inform him that his government wished to take into serious consideration the invitation to send a delegation to London. This invitation, which had also been addressed to Bulgaria and Hungary, was part of a more general forward policy towards Eastern countries in order to strengthen relations before the repercussions of the events in Vietnam and Rhodesia were felt. The fact that Maurer was now talking of sending one of the three First Deputy Prime Ministers – who had a higher rank than the Foreign Minister himself – automatically placed the visit on a different level from the simple exercise of diplomatic contacts as it was originally seen by the Foreign and Commonwealth Office (FCO) and, to make his point, the Romanian Prime Minister recalled his recent

111 UKNA, FCO 28/247: Rendall to McMeekin, 03.03.67; Chadwick to Brown, 29.02.68.

112 ANR, CC, Chancellery, 1968_22: notă cu privire la situația actuală a parcului de avioane pentru trasportul aerian internațional, 10.02.68, ff. 8-11. The lack of thrust reverser was confirmed by an article translated from a Soviet review: "Features of the Tu-134", *Flight International*, August 22, 1968, p. 288-290.

hunting expedition with the head of Krupp, Berthold Beitz. Romania was interested in talking business and furthering economic cooperation, initiating contacts with the main industries in the United Kingdom.[113]

Although filtered through the lens of ideology, the evaluation of the British position made by Romania was essentially correct. In the face of an attitude oscillating between indifference and hostility, Bucharest's anti-hegemonic policy had led to a change of course, pushing the United Kingdom towards a more active and open approach. Similarly, considering that the public debate was on the decline, the choice to play the card of emulation with the Germans was more than adequate and functional to the Romanian interests. The purpose of the visit, in fact, was to enhance trade, in particular by increasing the Romanian quota of agricultural and oil production and by making agreements that would have resulted in machinery and industrial plant useful to the realisation of the five-year plan; everything would have to find fulfilment in an agreement of bilateral scientific-technical cooperation. The interest from British companies was not lacking: BAC, HSA, Vickers, ICI, the electromechanical holding AEI and many others made evident to their governments their desire to get in touch with the Romanian delegation during the visit. In Whitehall, however, there was a tendency to regard the operation substantially in politico-promotional terms and to look at Romania with a certain caution, something that its leaders were aware of.[114]

It is interesting to note how this inconsistency still existed decades later. Sergiu Celac and Andrew Bache occupied similar positions in their respective Foreign Ministries and, in the early 1990s, they would find themselves, both with the rank of Ambassador, representing their countries in each other's capitals. While Celac remembers the unusual importance given to this visit and its non-propagandist character,[115] Bache's version of events goes in a different direction and has a very different connotation: "They wanted to have some suggestion of a Western link. [...] The

[113] UKNA, PREM 13/1172: Fenn to Wright, 21.12.65; UKNA, BT 11/6747: Glass to Smith, 17.12.65. On Beitz and diplomacy as German industrialists' 'hobby' in mid 1960s, see Norbert Pötzl, *Beitz. Eine deutsche Geschichte*, Munich, Heyne, 2011, p. 215-246.

[114] ANR, CC, Chancellery, 1966_7: notă privind poziția părții române față de problemele care se vor ridica cu ocazia vizitei în Anglia, 22.01. 66, ff. 15-29; UKNA, BT 11/6747: Crowe to McMeekin, January 28, 1966; UKNA, FO 625/14: Smith to Glass, desp. No. 15, 18.02.66.

[115] Sergiu Celac to the author, 29.03.2015.

Romanians didn't have any money but they were always prepared to pay with tomatoes. It was almost a case of tomatoes for anything".[116]

For a while, the British remained on tenterhooks, failing to get from the Romanian authorities the list of personalities who would make up the delegation. The only certain thing was that it would be led by First Deputy Prime Minister Alexandru Bârlădeanu, one of the architects of the evolution of Romanian politics under Gheorghiu-Dej and the main opponent to the reform of the CMEA as his country's representative in that organisation. The uncertainty stemmed from the fact that the Central Committee of the PCR had not yet deliberated on the instructions to give to the delegation. In any case, Bârlădeanu himself had made it clear, beyond any possibility of misunderstanding, what the purpose of the visit was, namely, the importance of activating economic contacts and the intensification of bilateral relations. This additional indication was useful in the sense that the Board of Trade reached, at the very least, the conclusion that the British government would have to show interest in the strengthening of technical and industrial cooperation, asking the Romanians for concrete proposals that, later on, individual traders would take into consideration.[117]

On January 24, 1966, the Permanent Presidium of the Central Committee approved the instructions for the delegation. To begin with, a target quantity for the increase in trade was set, but the main purpose was to point out that, as far as the British were concerned, there were other issues at stake. On the one hand, it was necessary to provide for an increase in Romanian exports so as to offset the value of technological imports and industrial equipment; on the other hand, it was essential to insist that the quality of these imports reached world standards, given that Romania was not obliged to purchase from the United Kingdom. Had the British raised the issue of the possible acquisition of aircraft, the delegation would have had to answer that there was no interest in this for the time being.[118] Given that conversations were already in progress, it is highly likely that this arrangement had been motivated by the fact that no choice had yet been made, either of a technical or economic nature,

---

116 Andrew Philip Foley Bache, British Diplomatic Oral History Programme, available at www.chu.cam.ac.uk (last accessed on July 29, 2015).

117 ANR, CC, Chancellery, 1966_7: notă de audienţa, 21.01.66, ff. 13-14; UKNA, BT 11/6747: brief on joint venture, 27.01.66; UKNA, FO 625/14: Glass to FO, tel. 28, 20.01.66.

118 ANR, CC, Chancellery, 1966_7: protocol No. 3 al şedinţei Prezidiului Permanent al C.C. al P.C.R., 24.01.66, ff. 2-6.

on the type of aircraft to be purchased and that no one wanted to put too many irons in the fire, when the objectives of the visit were, essentially, to discuss only the technological cooperation agreement and the issue of nuclear technology.[119]

The very composition of the delegation was a clear demonstration of the importance attached to the visit by the Romanians. In addition to Bârlădeanu, the delegation comprised Gheorghe Cioară, Minister of Foreign Trade, George Macovescu, Deputy Foreign Minister, and Vasile Pungan, who, shortly afterwards, was to become Ambassador in London and, later on, one of the closest associates of Ceaușescu. In a real *tour de force*, between February 2 and February 9, 1966, the Romanians visited several British businesses and they had meetings with Prime Minister Wilson, Foreign Secretary Michael Stewart, and the Chairman of the Board of Trade, Douglas Jay. Given the explicit purpose of the operation, it is not surprising that the more substantial talks were the ones held with the Board of Trade, which was keen to address the aircraft issue, in particular the offer made by HSA for the Tridents, considering that, if the deal had gone ahead, it would have been a decisive step forward for British penetration into the markets behind the Iron Curtain. Moreover, the limits of expansion for a country that did not have large reserves of precious metals or oil were obvious, and so was the risk of the accumulation of junk debt in the long run, especially when compared with a development based on the importation of machinery and technology, something which the French ambassador in Bucharest soon became aware of, alerting his government.[120]

For their part, Jay and the Minister of State Roy Mason pleaded the HSA cause, praising the technical superiority of the Trident against the French Caravelle, but the main argument was, supposedly, that the British seemed less flexible than the FRG, France and Italy in the face of the urgent Romanian request to expand the possibilities for counter-trade. Pressed by Glass, Bârlădeanu pointed out that, for example, the French absorbed about 20% of Romanian exports of petrochemicals – referring, in all likelihood, to the contribution made by the French government to the negotiation for the production of Renault cars in Romania – and that they no longer believed that his country was still an agricultural

[119] On the groundbreaking role of UK nuclear exports to Romania, see Eliza Gheorghe, "Atomic Maverick: Romania's negotiations for nuclear technology, 1964-1970", *Cold War History*, Vol. 13, No. 3 (2013), p. 373-392.

[120] UKNA, BT 11/6747: brief for the President's meeting with Bârlădeanu, n.d.; minute by Welch, 02.02.66; Archives du Ministère des affaires étrangères (MAEF), série 29, sous-série 29, dos. 296: Pons à Debré, No. 392/EU, 20.12.68.

one. Cioară, following the instructions given by the Presidium, told the British that, when negotiating contracts with them, it was necessary to add warranty clauses to guarantee performance, giving some concrete examples of the problems previously encountered. According to the Minister for Foreign Trade, the United Kingdom would have to do as the other Western partners of Romania had done, namely, draw up trade agreements that could establish long-term patterns of trade for several years, according to the organisational needs of a planned economy. When asked about this, the Commercial Advisor in Bucharest pointed out that Romania needed similar agreements with France and the FRG as a context in which they could exert major pressure for the sale of their own products by luring industrial equipment manufacturers with tempting orders.[121]

The meeting with Wilson was entirely conventional,[122] whereas during his briefing with Stewart, Bârlădeanu launched into a eulogy of British science and technology, and suggested sending a special task force to the United Kingdom to negotiate a cultural agreement: "Mr. Bârlădeanu commented that the British side were usually more interested in the humanistic aspect, whereas the Romanians preferred to emphasise the scientific and technical aspects. He would like the British Government to consider the matter and accept an agreement at government level stressing this aspect".[123] Significantly, there was yet again no talk of aircraft, even during the meeting at MinTech, where the main topics discussed were nuclear power, computers and electronics.[124]

Overall, the visit was a success, even on a personal level for Bârlădeanu, who corroborated the image of a country that was indeed Communist, but also dynamic and pragmatic:

> Mr. Bârlădeanu made a good impression. In formal discussions he spoke intelligently and with authority. His manner was conciliatory, even when his position was firm. On social occasions he was friendly and at ease. He and the majority of his colleagues did not present the stiff flat face, which is still so often the mark of the East European Communist. But then, as they are so fond of reminding us, they are Latin.[125]

[121] UKNA, BT 11/6747: Anglo-Rumanian Trade, minute No. 87, 04.02.66; note of a meeting at the BOT on 04.01.66; Randall to McMeekin, 04.03.66.

[122] UKNA, PREM 13/1172: record of a meeting between the PM and Mr. Bârlădeanu on 09.02.66.

[123] UKNA, PREM 13/3461: record of a meeting between the Foreign Secretary and Mr. Bârlădeanu on 04.02.66.

[124] UKNA, BT 11/6747: visit of Mr. Bârlădeanu, 14.02.66.

[125] UKNA, FO 625/14: Smith to Glass, No. 15, 18.02.66.

However, the limited relevance of the aeronautical issue in the British government documents is quite remarkable, confined as it was to some brief notations made by the Board of Trade, which was, on top of it all, entirely focused on the HSA offer. Moreover, even during the visit of Foreign Minister Mănescu to his new British counterpart, George Brown, in October 1966, the former went back to the Romanian interest in nuclear power and, more generally, to the difficulties of developing bilateral economic relations, without addressing the issue of the aircraft for TAROM.[126]

Even for the Romanians the visit had been a success. They believed that it could serve as a starting point for a real change of pace in bilateral relations, with a definitive negotiated solution to the issue of British financial claims, to the technological cooperation agreement and to the expansion of bilateral trade. The Presidium ordered that the economic organisations should try to implement the necessary measures to improve the range, quality and presentation of the goods based on the needs of the United Kingdom market in order to take full advantage of the existing quotas and to study the prospects for a future, long-term trade agreement. A new mission to the United Kingdom was also planned in order to finalise the agreement for technical cooperation. So, in early 1967, the President of the National Council for Scientific Research, Roman Moldovan, arrived in London with a delegation that was initially intended also to include Elena Ceaușescu: the agreement was signed and the dialogue with MinTech on possible cooperation in the production of computers went on.[127]

The Presidium also ordered that the Ministry for Foreign Trade cultivate relationships with those companies willing to cooperate with Romania and it was in this context that the issue of the aircraft could move forward. In January 1967, BAC renewed its offer and, in February, Cioară went to France to visit Sud Aviation – the manufacturer of the Caravelle – but it was only in March that Tehnoimport, the Romanian state agency responsible for foreign trade in the technology sector, was authorised to actively begin the purchasing procedures. The negotiations would prove difficult not only due to a highly centralised system like the Romanian one, where the main negotiators involved depended on the

[126] UKNA, PREM 13/3461: record of a meeting between the Foreign Secretary and the Rumanian Foreign Minister on 10.10.66.

[127] ANR, CC, Chancellery, 1966_17: protocol No. 7 al ședinței Prezidiului Permanent al C.C. al P.C.R., 16.02.66, ff. 2-9; UKNA, FCO 28/255: Bucharest to FCO, tel. 24, 17.02.67; Bucharest to FCO, tel. 36, 25.02.67; "Rumanian Interest in Computers", *The Times*, March 13, 1967.

supreme decisions of the party, but also because it became increasingly clear that Romania's objectives were getting more complex. In fact, the aim now was not simply the purchase of modern aircraft for the transport of tourists, planes that otherwise would have to be hired, nor the mere achievement of its commercial compensation to cover the expenditure of hard currency. The real purpose was using this important transaction to create a relationship of technological and industrial cooperation thanks to which the aviation industry in Romania could be re-launched. In July 1967, Cioară told the British ambassador John Chadwick that TAROM's future supplier would have to agree to set up joint production of light aircraft in his country under licence, in order to cover part of the payment due.[128]

Such an approach quickly eliminated HSA, who were not willing to guarantee this kind of cooperation, and who, in any case, were unable to offer counter-trade, apart from a rather small percentage of the contract price. Instead, prompted by the Commercial Advisor in Bucharest, Archie Rendall, BAC actively took into account Romania's demands, quickly becoming the British frontrunner. For its part, Sud Aviation, which in the second half of 1967 was still considered the favourite, sent its final offer in October via a letter penned by its president, the infamous Maurice Papon. In that letter, Papon envisioned a possible cooperation on three aircraft models: the Rallye, a small single-engine propeller plane that could be used for tourism, training or as a utility aircraft; the Paris III, a twin-engine jet with features that corresponded in terms of performance and possible use for the requirements – military and, if need be, civilian – highlighted by Romanian studies, but whose production remained suspended pending the materialisation of a sufficiently substantial initial order on the part of the French government; and the Potez 842, a business aircraft powered by four turboprops, which was of no interest to the Romanians because of its high costs. Even in the case of Sud Aviation, although cooperation proposals were really significant, the bulk of the cost of the Caravelles still remained uncovered.[129]

On August 4, 1967, BAC's regional sales manager, Anthony Rhoades, sent the company representative for Romania, George Pop, to Bucharest,

---

[128] UKNA, FCO 28/247: Rendall to McMeekin, 23.03.67; Chadwick to Brown, 29.02.68; Chadwick to FCO, tel. 197, 18.07.67.

[129] ANR, CC, Chancellery, 1968_22: notă cu privire la situația actuală a parcului de avioane pentru trasportul aerian internațional, 10.02.68, ff. 12-13; Papon to Cîmpeanu, 16.10.67, ff. 17-19; UKNA, FCO 28/247: minute by Rendall, 05.05.67; Chadwick to FCO, tel. 136, 22.07.67; minute by Golden, 24.07.67; minute by Chadwick, 10.08.67.

with an offer to set up in the country full production of the Britten-Norman (BN) Islander, a twin-engine, nine-seater piston aircraft, newly certified and likely to be very successful. Pop, a Romanian who later became a UK naturalised citizen, would play a key role in all the decades of Anglo-Romanian cooperation as representative and problem solver for both BAC and Rolls-Royce. A staunch supporter of the revival of the aviation industry in his country as a symbol of national greatness and source of development, he devoted unremitting efforts to the affirmation of British products, activating a formidable network of contacts involving the highest levels of leadership in Romania, in a process of mediation that summed up his attachment to both his country of origin and to that of adoption.[130]

Brought to Bucharest on a demonstration flight by John Britten, the Islander made a favourable impression on the Romanians, magnetising their attention and basically offering them what they wanted, that is, the opportunity to gain experience with the installation of a modern but simple aircraft that would then be marketed directly by Britten-Norman. In addition, together with the Foreign and Commonwealth Office and the Ministry of Technology, the offer – which contemplated the supply of assembly kits to Romania and the withdrawal of 'green' aircraft for their final customisation and sale, an arrangement not to be submitted to CoCom.[131]

With Romania's final decision expected by the end of October, Pop took the British diplomats by surprise, recommending that BAC should invite Deputy Prime Minister Verdeţ to the United Kingdom. MinTech and the FCO initially opposed this completely independent initiative, lest it was based on wrong information. The commercial attaché in Bucharest, John Birch, asked Pop why, for example, he had thought of Verdeţ and not Bârlădeanu. The BAC representative retorted that he had a friend who, in turn, was friends with the secretary of Verdeţ himself, having thus had the opportunity to ascertain that the aeronautical dossier was in his hands; therefore, he had considered the invitation beneficial in terms of a psychological ploy, mainly designed to concentrate the Deputy Prime Minister's attention on BAC, precisely at a time when the agreement was most expected. Verdeţ was one of Ceauşescu's younger collaborators and

[130] UKNA, HO 334/408: certificate of naturalisation No. BNA46760. Celac to the author, 29.03.2015; Stanley Hooker, *Not Much of an Engineer. An Autobiography*, Marlborough, Airlife, 2002, p. 199.

[131] UKNA, FCO 28/247: Chadwick to Rhodes, 31.08.67; Rhodes to Chadwick, 18.09.67.

one whom the Secretary General was trying to raise in rank, as his control over the Party, and therefore his power, was progressively consolidating. If one considers that the British Embassy received specific indications from Czechoslovakia about an imminent changing of the guard only at the end of November, Pop had been able to infer the beginning of the end of Ceauşescu's collaboration with Gheorghiu-Dej's men, besides being able to foresee the closing of the season of limited reforms, well before the official diplomatic service could realise it, and he was thus able to act accordingly. For Whitehall, there was now no other option but to join in the game.[132]

On October 19, 1967, Ambassador Chadwick himself handed Verdeţ an invitation from the Vice-president of the Weybridge Division of BAC, Geoffrey Tuttle, asking him to visit the company, along with an express request from Rolls-Royce to visit their engine manufacturing division in Derby. Verdeţ declared himself very happy about the invitation, which he would willingly take up towards the end of the year, when his commitments allowed him to do so, adding that the proposals for cooperation with BAC were excellent, and that the Moldovan mission had shown good prospects in the field of computers and electronics. Thus, Tuttle's invitation was followed by one from Benn, this time to discuss this field of industry, considered to be the jewel in the crown by MinTech. Chadwick concluded that Verdeţ's visit would be a coup for British diplomacy, because it would be the first public appearance of one of the new men at the top of the Romanian *nomenklatura* outside the Communist world, with the exception of Finland. Moreover, believing that Ceauşescu was not going to take over the de jure and de facto leadership of the country, Chadwick predicted that in the long run Verdeţ himself could emerge as Romania's leading man.[133]

In late November, everything seemed to be in question again when an information leak revealed that the Romanians were ready to finalise negotiations, but Soviet pressure on Bucharest for the purchase of the Tu-134 was increasing. As the Deputy Minister for Foreign Trade, Ion Cîmpeanu, explained to Chadwick, the USSR was now offering six planes whose cost would be fully covered by the purchase of locomotives manufactured in Romania. Although the package offered by BAC was good, there still remained a part to be paid in foreign currency, which

132 UKNA, FCO 28/249: Rhodes to Hanbury-Tenison, 09.10.67; Birch a Rhodes, 18.10.67; Golden to Birch, 23.10.67; Chadwick to Smith, 23.11.67.

133 UKNA, FCO 28/249: Tuttle to Verdeţ, 19.10.67; Chadwick to FO, tel. 188, 20.10.67; Chadwick to Smith, 15.11.67.

created difficulties for the Romanian government, not to mention the fact that the risk of buying in the West could push Moscow to resort to economic pressure against Romanian exports. It would have been different, continued Cîmpeanu, if BAC had introduced a programme of long-term cooperation that would offer not only new sales opportunities for Romanian products, but also a partnership aimed at reviving the domestic aviation industry. In this context, Verdeţ's visit – finally scheduled between January 23 and February 3, 1968 – took on an even greater importance than originally thought.[134]

The package offered by BAC could hardly be improved from the financial point of view. In contrast to a total of 11.7 million pounds (of which 9.2 million for the six 1-11s and the rest for components and spare parts), netting arrangements would have to generate revenue of just under 8 million pounds for the Romanians. The difference would be covered by a credit of 5.5% with a maturity of seven years. An assembly line for the production of 215 Islanders over a five-year period would be set up in Băneasa: in theory, these aircraft would have a lower cost than those produced by Britten-Norman, but quality control represented an unknown variable for which, if necessary, BAC would have to take responsibility. The latter had involved Barker Machine Tools & Equipment Co., a Middlesex-based enterprise which, in exchange for the exclusive for the United Kingdom and some Commonwealth countries, had agreed to purchase a significant equivalent in machine tools produced in Romania – provided that BAC agreed beforehand to buy out a large proportion of it. Finally, the Zurich company Waltrade A.C. would buy Romanian goods for more than five million pounds to be sold on to other European and African countries. BAC even managed to convince a British travel agency to use the TAROM 1-11s for charter flights during the summer season. For its part, the British government, which had just suffered the shock of the devaluation of the pound, was not particularly keen on financially supporting transactions related to the aviation industry, then viewed with suspicion. On the contrary, the President of the Board of Trade, Anthony Crossland, would insist with Verdeţ that the trade balance be redressed favourably for the United Kingdom, given the deplorable state of its financial situation.[135]

---

134 UKNA, FCO 28/247: Chadwick to FCO, tel. 226, 23.11.67; Chadwick to FCO, tel. 250, 19.12.67; UKNA, FCO 28/249: Bucharest to FO, tel. 338, 30.11.67.

135 UKNA, FCO 28/249: minute by Randall, 03.11.67; trade with Romania, private office minute No. 88, 29.01.68; UKNA, FCO 28/251: sale of BAC 1-11 aircraft, 18.01.68.

From BAC's initial invitation onwards, Verdeţ's visit turned into a major event of about ten days, from January 23 to February 3, 1968, with a delegation that included the Deputy Foreign Minister, the Deputy Minister for Foreign Trade, the Secretary General of the Ministry of Transport, Constantin Şendrea (who, up to 1966, had been the general manager of TAROM) and the Dean of the Electronics Polytechnic of Bucharest, Mihai Drăgănescu, now Head of the State Committee on Computer Technology, along with twenty experts representing as many industries. Apart from a few moments of embarrassment, such as when the cars made available to Verdeţ broke down twice, the visit was a success, since it was agreed that a long-term business cooperation aimed at developing industrial Romania was in the mutual interest of both parties. From the British perspective, this meant trying to take advantage of a time when relations between Bucharest and Moscow seemed to be wobbling, and to try and change the flow of trade in a manner that would be favourable to London's balance of trade. The specific economic situation of the United Kingdom, therefore, contributed to concentrating the government's attention on relations with Romania. Although the immediate aim was to leave no stone unturned to correct the decline, this closer relationship could be useful even on a more general level, for instance, in the field of international affairs.[136]

The visit was a success for the government, in particular thanks to the signing of an agreement with Elliott Automation to cooperate in the creation of a computer industry in Romania, a priority project for MinTech, which was working at that time on the creation of a national champion in the field,[137] and also thanks to the establishment of joint working groups to examine the individual industrial sectors with a view to a future long-term trade agreement. Conversely, it was a disappointment for BAC because, despite all the pressure exerted, not even the heads of agreement were signed. In reality, the signals were all very positive. Verdeţ confirmed that there had been no doubt about the 1-11 from a technical standpoint, and he admitted that the package of counter-trade could not be further improved. However, as he explained to Wilson, in Bucharest the Islander component was still being discussed and, above all, the aim was now to lay the foundations for the revival of the aviation industry in Romania. In this regard, it was very important to be able to cooperate in

---

136 UKNA, FCO 28/249: Birch to Goulden, 03.01.68; UKNA, FCO 28/250: Brown to Chadwick, 10.02.68; Chadwick to Tom Barker, 08.02.68.

137 Martin Campbell-Kelly, *ICL. A Business and Technical History*, Oxford, Clarendon, 1989.

the development and production of the BAC 201, still at the draft stage; the aircraft would be the perfect choice for serving Romania's domestic routes. Verdeţ declared himself positively impressed by the BAC package as a whole and promised a reply from his government by the end of the month.[138]

It is hard to underestimate the importance of Verdeţ's message, whose positive impression was, in all likelihood, the element that definitively tipped the scales in favour of the British. Beyond the possible economic pressure from the Soviets and the obvious advantages in terms of exchange, the purchase of the Tu-154 was not a viable option, because Moscow communicated in early 1968 that it could not have delivered the planes before 1970 and that, in any case, that model would be produced in a small series to be quickly replaced by another. Now, if the British and French offers were basically the same in terms of the currency effort, the BAC package offered larger-scale and more binding commitments in terms of counter-trade – a very important aspect since, as noted by the State Planning Commission, for 1968 the five-year plan had not budgeted funds in hard currency to buy planes in the West, and so it was necessary that the initial cash flows were fully compensated. Moreover, while the production of the Rallye would require substantial investments without the plane finding a certain use in Romania, the assembly of the Islander would be inexpensive, and the 215 planned units had been judged sufficient to generate profit. Above all, it was a model more suitable for Romanian development plans – so much so that its technological assimilation had already been foreseen. Finally, BAC had hinted that if TAROM bought the 1-11, it would be possible to ask the British authorities for the right to land at Heathrow, rather than at the less centralised Gatwick Airport, with a relative increase in prestige and, probably, in the number of passengers flying with the Romanian state company.[139]

The most crucial element, however, was the prospect of collaborating in the realisation of the 201, together with the availability expressed by Rolls-Royce to produce the engines in Romania. The 201 was a simple plane, without pressurisation and with a fixed undercarriage, designed to be inexpensive and to be able to shift easily from passenger transport (with up to 28 seats) to freight (9,000 lb.) It was powered by two Dart

---

138 UKNA, PREM 13/2361: record of a meeting between the PM and Mr. Verdeţ on 02.02.68; UKNA, FCO 28/250: FO to Bucharest, tel. 42, 03.02.68.

139 ANR, CC, Chancellery, 1968_22: notă suplimentară privind dotarea aviaţiei civile cu avioane destinate transportutilor aeriene internaţionale în perioada 1968-1970, n.d., ff. 21-25; Pîrvu to Baicu, No. 64396, 09.02.68, ff. 26-29.

turboprops, engines designed twenty years before with the old centrifugal compressor technology, rather than with the latest axial compressors, and it had experienced wide diffusion and longevity.[140] It was precisely the type of plane that was deemed essential for the recovery of the aviation sector in Romania, a machine that was small enough and relatively simple, whose engines would be produced under licence and that could be used both for civil and for military transport. In addition, the long-term cooperation with the British industry in this field would make it possible to verify the possibility of extending it to the production of fighter aircraft. Here, the attention of the Romanians was focused on the French Mirage F3 and on the Lightning F Mk-3, produced by English Electric and then by BAC and sold to the Saudis and Kuwaitis in the 1960s. The licensed production of one of these aircraft would have allowed the Romanian industry to assimilate the necessary know-how to later develop them independently starting from the mid-1970s.[141]

Therefore, it is perfectly understandable why the cost-effectiveness of the British offer, while duly emphasised by Bârlădeanu, was not the only reason for the choice in favour of the 1-11, nor – in Ceaușescu's view – was it merely an attempt to give TAROM modern equipment. During a meeting of the Executive Committee of the PRC on February 15, 1968,[142] the Secretary General recalled the request recently made to the Soviets to obtain the licence for the MiG-21, pointing out that it was all but certain that Moscow's response would be swift and positive.

> Comrade Nicolae Ceaușescu: 'We, therefore, approve the purchase of six BAC 1-11 aircraft. [...] Our concern is to produce small planes having up to 50 seats and to begin discussions for small jet engines for military aircraft. For the time being, conversations should take place unofficially, but let's start them, because we talk about these issues with the Soviets, but they might just as well tell us that they will give us what we want in ten years' time.'[143]

---

140 "Paris: Biggest Ever", *Flight International*, 01.06.67, p. 899; Peter Pugh, *The Magic of a Name: The Rolls-Royce Story*, Vol. 2: *The Power Behind the Jets, 1945-1987*, Duxford, Icon Books, 2001, p. 36-44.

141 ANR, CC, Chancellery, 1968_29: studiile preliminare privind stabilirea de Industria Aeronautică Română, n.d., ff. 38-59.

142 The Executive Committee took over the functions of the Politburo of the Party in 1965. See Nicoleta Ionescu-Gură, "Studiu introductive", in Florica Dobre (ed.), *Membrii C.C. al P.C.R. 1945-1989. Dicționar*, Bucharest, Ed. Enciclopedică, 2004, p. 11.

143 ANR, CC, Chancellery, 1968_29: stenogramă ședinței Comitetului Executiv al C.C. al P.R.C., 15.02.68, f. 3; Petre Opriș, „Colaborarea lui Ceaușescu cu British Aerospace", *Jurnalul Național*, 17.08.09.

It was no coincidence, then, that in the following days the Romanian negotiators insisted on the creation of a joint venture for the progressive installation in their country of the production of the 201 – or, at least, of a comparable model, should the plane not obtain the necessary funding for the launch from the British government, including the Rolls-Royce engines. BAC, actually, was not willing to take on binding commitments in this regard, so on February 22, when the agreement for the purchase of the six 1-11s was finally signed, this aspect was regulated in a separate memorandum, in which it was simply specified that the utmost consideration would be given to Romania as a partner, provided that the 201 project entered production and that all the parties involved were satisfied with regard to technical proficiency.[144]

For the Romanians, it was a very ambitious programme, in which the purchase of the BAC planes only served to trigger a process of modernisation of the national mechanical industry. They would acquire the ability to produce sophisticated military equipment, thus proving their capacity to sustain an intensive growth, also quality-wise, in a leading technology sector. Sometimes the line between audacity and temerity is not easily distinguishable and, in fact, Maurer tried to challenge a programme that aimed, first and foremost, to give Romania an independent production capacity in military aviation. Little wonder that the Executive Committee did not support him. The Minister of the Machine-Building Industry, Mihai Marinescu, reminded him of how the aviation industry in Romania had started precisely with military production in the period between the two wars; the Defence Minister, Ion Ioniță, pointed out that they were not starting from scratch, because there were already many specialists in the country engaged in the maintenance of Soviet-built aircraft, and that Romania was already producing a large proportion of their components. Ceaușescu, revealing how power management was undergoing significant changes in the PCR, cut things short by pointing out that he was not merely talking about military aircraft, but about a broader strategy, to be fully implemented in stages: firstly, the assembly of the Islander and the production of an aircraft with 30-50 seats like the BAC 201 that could be used for both civilian and military purposes, all to be realised with the help of the British, thus contributing to the progressive organisation of military production. Maurer, Ceaușescu concluded, should stop worrying unnecessarily, because the prospective Romanian fighter aircraft would not be put together independently, but with the help of the country most

144 UKNA, FCO 28/247: Chadwick to FCO, tel. 12, 16.02.67; FCO to Bucharest, tel. 65, 16.02.68; minute by Rendall, 26.02.68.

willing to cooperate – be it the USSR or another. At this point, it was a matter of little importance, and besides, even advanced countries like the United States and France resorted to cooperation in the aeronautical sector.[145]

For its part, the British government, however, did not seem to fully understand the political dynamics in place in Romania, nor did it grasp the importance of the assignment of a real priority to transactions in the aeronautical sector. The pages that Benn dedicated to his visit to Bucharest between June 3 and June 8, 1968, are in some ways surprising, not so much for their focus on the Anglo-Romanian cooperation on computers, marked as always by the competition with France and by an obvious impatience for the mechanism of CoCom, but rather for the idea that modernisation was a process capable of determining an evolution of the Romanian regime towards greater political openness and that technological cooperation could act as a catalyst for this change. Obviously, this process would have important effects on Détente and, ultimately, on the ending of the Cold War: "[…] I thought that our two systems were like a piece of cheese and a piece of butter both going round in a soup which technology was making hotter and hotter; that in this technological soup, what would come out would not be butter or cheese but something perhaps between the two".[146] The diplomats who did not share this point of view were accused, not even too implicitly, of showing an anti-Communist prejudice, of having a Tory and imperialistic vision of international relations, a view that prevented them from understanding the evolution of Romania, although at least Ambassador Chadwick echoed Benn when he noted that the technological interlocking between the two countries could actually influence the long-term evolution of the Romanian regime, in a way that could end up being beneficial to British interests as well.[147]

This kind of approach, together with the result of the dispute over the fate of the domestic aviation industry in the United Kingdom, did not help in understanding the reasons and the goals of Bucharest in the development of aviation. Whereas the President of the Commission of Technical and Economic Cooperation, Gheorghe Rădulescu, insisted on how important it was for his government to participate fully in the 201 project – about which the negotiations continued, with BAC open to a Romanian participation of up to 30% in the design and up to 75% in

145 Petre Opriş, *Industria românească de apărare: documente, 1950-1989*, Ploieşti, Ed. Universităţii Petrol-Gaze, doc. 48, p. 244-250.

146 Antony Benn, *Office Without Power. Diaries 1968-72*, London, Arrow, 1989, p. 78.

147 UKNA, FCO 28/257: Chadwick to Stewart, 19.06.68.

the production phases –,[148] on the other hand, Benn emphasised that perhaps, for a country like Romania, it was more appropriate to focus on components:

> The Minister went on to emphasise that while he appreciated Romania's desire for independence in certain fields, it was almost impossible, with such limited resources, to be completely independent in aircraft manufacture. It might be that Romania's future in the aircraft industry would be as a supplier of components rather than as the manufacturer of complete aircraft.[149]

Benn could, after all, be right, but the prospect that he was pointing to was very different from the one decided by the Romanian leaders; likewise, the wisdom of teaching Ceaușescu a lesson on the importance of market mechanisms in order to decide whether to proceed with the BAC 201 project or not, although perhaps a moment of frank sincerity, is a matter of debate.[150] The vicissitudes of the aviation industry in the United Kingdom and the consequences of the Prague Spring would further complicate the prospects of Anglo-Romanian cooperation.

---

148 UKNA, FCO 28/257: note of a meeting held on 04.06.68 at 9:00 a.m. corrigendum.

149 UKNA, FCO 28/257: note of a meeting held on 05.06.68.

150 UKNA, FCO 28/257: note of a meeting between Benn, Ceaușescu and Maurer on 07.06.68.

CHAPTER 2

Counter-trade: the B-N Islander Production Line in Romania

The installation of an Islander production line in Romania was not only an important element of counter-trade, but it also gave the national industry an opportunity to cut their teeth on a plane technically within their reach and which had an internal, as well as an external, market. Above all, the Islander provided the opportunity to learn how to produce according to Western quality standards, which – in principle – the Romanian industry was able to do when one considers that both public testimony and internal documents attest to a high level of workmanship quickly reached.

Unfortunately, due to unlucky circumstances, assessment errors and the weakness of the British aviation industry, Britten-Norman (BN) had the bad habit of failing. In both of its major bankruptcies, the Băneasa line survived because it was now firmly at the crossroads of almost all Anglo-Romanian relations in the aviation industry. If anything, a relative lack of transparency on the part of the British partner of the moment should be noted, focused as they were on their own interests and, ultimately, on getting money from Romania.

2.1 The Vagaries of Insolvency

Counter-trade was the essential element and the objective necessity of any East-West transaction that aspired to sustainability in the medium-to-long term, taking into account the exchange constraints and the resources available. This was certainly true for the sale of the six BAC 1-11s to TAROM that, as has been established in the previous chapter, was the centre of a complex package which contained much of the significant prospects for cooperation – thus lowering the costs of the learning curve for the aviation industry in Romania, as well as providing a number of compensatory actions in Romania's favour. Among the latter, the most significant was the subcontracting, by BN, for the production of 215

Islander aircraft in favour of the Enterprise for the Repair of Aeronautical Material (Întreprinderea de Reparatii Material Aeronautic, IRMA) at the Băneasa airport, close to Bucharest.

In 1951, Britten and Desmond Norman, after an initial training under de Havilland, founded their company in Bembridge, on the Isle of Wight. At the time, BN acquired the old Tiger Moth training biplanes, produced by de Havilland, which the RAF was phasing out, in order to turn them to agricultural use and later sell them successfully in New Zealand. In 1960, Britten and Norman created another company, Cameroons Air Transport Ltd., which operated a regular air service across the estuary of the Mungo River, joining Douala and Tiko. It was thanks to this activity that they realised that on the market there were no sturdy, cheap aeroplanes combining a high seating capacity with the possibility of converting the entire cabin space for quick goods transportation. Thus, in the course of 1965, Britten designed and built the first Islander prototype entirely with his own funds. If BN originally thought about a small-series production, favourable market prospects – especially those presented by its distributor in the United States, Jonas Aircraft – pushed the company to pursue more ambitious programmes and apply to the government for funding.[1]

The Islander was designed to have a good STOL (short take-off and landing) performance and a structural strength enabling it to be used on rather bumpy runways, while remaining an inexpensive plane. To obtain this result, Britten had chosen conventional technical solutions and production methods, so as to minimise the capital cost for each single saleable seat. For example, sheet metal parts were produced simply by means of a router and a rubber-die press, available in a normal workshop of fine mechanics; similarly, Britten planned an intensive use of lamination for the production of spars, stringers and inter-spar plating, both to limit machining costs, and to facilitate repairs.[2] This meant that Romania could usefully participate in the production of the aircraft, without expensive investments in special machine tools – which would have posed a problem at CoCom level – starting from substantially traditional production practices.

Thus, the production of the Islander in Romania offered multiple simultaneous benefits. In fact, not only was it an important element of counter-trade, but it also provided an opportunity to develop a level of quality corresponding to Western standards from a product that did not

1 "Britten-Norman Background", *Flight International*, 28.10.71, p. 679.

2 T.E. Ford, "The Britten-Norman Islander", *Aircraft Engineering*, Vol. 42, No. 8 (1970), p. 16-23.

require too big a technological leap. Moreover, it was a type of machine that Romanian planners thought might be usefully employed in their own country for civil and military purposes, in line with existing production capacity, which was concentrated on gliders and small utility aircraft. It became feasible, in fact, to follow in the aviation industry the same strategy that was being adopted in cooperation with France's Renault in the field of cars: assembly of the R8 (Dacia 1100) starting from components imported from France and with the assistance of French technicians in order to develop appropriate productive skills and to participate in the R&D of a more complex model, the R12 (Dacia 1300); at a later time, the transfer of the R8 line production teams, already formed, on the R12 line.[3] A very similar course of action could be followed with British cooperation and, indeed, things worked well until the discrepancy between technological complexity and available resources became too wide. For example, the director of IRMA, Ion Trofin, was then sent to run Turbomecanica, an aircraft engine factory, and most of the highly trained personnel on the Islander line were transferred to activities relating to the BAC 1-11.

Paradoxically, difficulties would arise not from the Romanian side, but from the British side, and even before the final closing of the agreement with BN on May 6, 1968, that is, three months before obtaining the Public Transport Certificate of Airworthiness in the United Kingdom. The British company was having both financial and staffing problems, and it was unwilling to relinquish too many experienced engineers, to be sent to Băneasa for a relatively long period in order to give assistance to the production line in Romania, while also controlling its quality. This stemmed from the fact that, simultaneously, BN was undergoing a radical restructuring. The facilities in Bembridge were insufficient for the new production targets, so it was decided that BN would focus only on the final assembly of 'green' planes subcontracted by the British Hovercraft Corporation (BHC) at East Cowes, also on the Isle of Wight, and by IRMA at Băneasa. In this case, planes would be brought to Bembridge by ferry and all the avionics would be mounted there. The problem was that with tooling costs higher than expected, penalties for late deliveries, overtime wages and investments linked to the learning curve, BHC had already accrued one million pounds of losses by 1969, when the recession

3 Beatrice Scutaru, *Les relations entre les sociétés française et roumaine des années 1960 à 1995: un atout pour l'ancrage de la Roumanie à l'Europe?*, Ph.D. dissertation, Université d'Angers, 2013, p. 87-91.

in the United States and the arrival of new competitors reshaped much of the market outlook.[4]

In such a situation, the Romanian production could have played a very important commercial role, especially in the light of the very low unit cost of the aircraft coming from Băneasa (6200 pounds)[5] and the fact that the issue of return on investments was probably not the most urgent in a planned economy such as the Romanian one. The problem, however, was that, in spite of a very high standard of workmanship, the Islander assembled by IRMA could not be sold in the United States. In fact, since June 1968, the Federal Aviation Administration (FAA) had stated that it could issue a type certificate only for aircraft produced in a country that the United States had a separate bilateral agreement with, and Romania was not one of these. The FAA, then, wondered if there was any intention of exporting IRMA's products to its country, since, at that point, there would have been a problem. The pressure exerted by the Air Registration Board (ARB) on BN so that it clarified the situation with the Americans had no effect. In the first months of 1971, by which time the company was already going through a very rough period, the ARB checked with the FAA whether or not a more 'inclusive' interpretation of the Anglo-American bilateral agreement was possible. This would have included the Romanian Islanders, without explicitly assigning part of the production to a third country. The American response was totally negative, noting that, in their opinion, it was out of the question for the ARB to take legal responsibility for certifying the quality control of production in a country where it had no jurisdiction.[6]

BN's difficulties naturally reverberated on Romania, too. With about forty aircraft completed, the Romanians were interested in an extension of the original contract, but pointed out serious problems. The Deputy Minister who was in charge of international cooperation at the Ministry of Machine Building Industry, Nicolae Cocoş, and Stefan Ispaş, General Manager of the Division of Aircraft Manufacture, complained about the delay or non-delivery of materials by BN and about how some of these showed totally unacceptable levels of corrosion, believing that this was a deliberate attempt to slow the Romanian production on account of the

4 UKNA, FCO 28/248: minute by Rendall, 25.04.68; minute by Golden, 06.05.68; record of a meeting on June 15 between Knight and Verdeţ; "The Islander Saga", *Aircraft Engineering*, Vol. 42, No. 8 (1970), p. 6-14.

5 UKNA, FCO 28/247: minute by Rendall, 26.02.68.

6 UKNA, DR 8/30: Haldeman to Ellis, 05.06.68; Guilbert to Fenner, 30.07.70; Guilbert to Cary, 19.03.71; Fenner to Guilbert, 19.04.71.

difficulties in selling the product. Britten had to go to Bucharest personally to convince the Romanians not to initiate litigation, but it is significant that from the contacts with Mark Norman, Desmond's older brother, who was in charge of the financial and business matters of the company, the Department of Trade and Industry (DTI) gleaned the impression that BN had been less than transparent with its partners.[7]

The seriousness of the situation became apparent in the summer of 1971, when the auditing firm Price Waterhouse signalled the company's critical financial situation and the director of Exporters Refinance Corporation (ERC), a subsidiary of Lloyds Bank, who had been following the operations of credit in favour of BN, resigned. At that point, it was a race against time to find ways to operate an injection of fresh capital that would allow the company to survive. The most practical option was the sale of the licence for the production of the Trislander, a three piston-engine plane, which was an evolution of the Islander, to Israel, mostly through the mediation of the former Deputy Chief of Staff, Ezer Weizman, the future president of the country. Kannaf, an air taxi company, was ready to buy ten planes immediately; at the same time, the rights of production and distribution would be transferred to Israel.[8]

On September 21, the FCO informed the ambassadors in Bucharest and Tel Aviv of BN's troubling financial situation, directing them not to tell the respective governments. On the same day, BN's permanent representative in Romania, Kenneth Mills, informed the Commercial Counsellor, who three days earlier had brought to the attention of Ispaş the possibility of extending cooperation to the Trislander. Mills, in fact, believed that the agreement with the Israelis, of which he had been informed only recently, would compromise the sales prospects of the military version of the Islander in the Gulf states, which he considered very promising in the context of the tense situation in the Middle East. Consequently, he pressed BN's management to verify the feasibility of a Romanian solution, taking advantage of the fact that the signature of the Protocol with Israel, originally scheduled on September 25, had been postponed by a fortnight. Talks about the Trislander had been going on ever since March 1971, but only on September 18 did Mills and Tehnoimport discuss the possibility of establishing a joint venture, which, by exploiting Romania's international position, would be able to sell both

7 UKNA, FCO 28/1532: Russell to DTI, tel. 7, 29.01.71; Shand to Russell, 05.02.71; Russell to Shand, 16.03.71.

8 "Six Bids for Britten-Norman", *Flight International*, 11.11.71, p. 761-762.

to the Arabs and the Israelis. As a result, Britten's trip to Bucharest was hastily organised for the next day, September 22.[9]

According to the British ambassador to Romania, Denis Laskey, the fact that BN had entered into negotiations with the Romanians was a sign that the transaction with Israel was not to be taken for granted. Now, the situation was already difficult because the production of the Islander had been the most successful element of Anglo-Romanian aeronautical cooperation until that moment and, consequently, its absence would have represented a heavy setback, which would ultimately be attributed to the basic unsoundness of the British industry. In addition, BN now aimed to achieve an immediate loan of 2 million pounds from the Romanians in the hope of staying afloat, but without fully informing the counterpart of the real gravity of the situation. Although Laskey recommended maximum transparency, the departments involved at Whitehall soon reached the conclusion that the government had to keep away from this imbroglio to avoid socio-political gaffes and accusations of acting in bad faith. BN would have most likely gone bankrupt anyway, so there was no point in risk-taking, but rather it was important to organise a meeting with ERC and BAC, which, at that time, was negotiating a new transaction with Romania, to inform them of the situation. The priorities were to make sure, on the one hand, that the Romanians – who had confirmed their interest in the Trislander – would not feel cheated, and, on the other, that the position of BAC would not be compromised by BN's desperate attempts at survival.[10]

Apart from any political and security considerations about an operation that would involve a member of the Warsaw Pact becoming a partner in an aviation company based in one of NATO's most prominent countries, the idea that the Romanians could decide in favour of a financially important and risky investment in such a short time primarily came across as an index of BN's desperation, rather than as a manifestation of wishful thinking. Moreover, being convened to Whitehall to openly discuss the seriousness of the situation could hardly serve as an incentive for ERC to keep its credit lines open, considering that the company had already lost considerable amounts of money, together with Lloyds, in the spectacular failure of Rolls-Royce that took place in February 1971. In fact, on October 21,

9 UKNA, FCO 28/1532: minute by Russell, 23.03.71; Home to Bucharest and Tel Aviv, tel. 402, 21.09.71; Russell to Reed, 21.09.71; Russell to Hanbury Tenison, 28.09.71.

10 UKNA, FCO 28/1532: Laskey to FCO, tel. 308, 22.09.71; Laskey to FCO, tel. 321, 25.09.71; note of a meeting held on September 30, 1971; Home to Bucharest, tel. 432, 08.10.71; Russell to Downey, 12.10.71.

Lloyds made it known that the next day it would call for the appointment of a receiver for BN. The Cabinet, which met on the same day, examined the situation: the Minister of Aerospace, Frederick Cornfield, noted the deleterious effect of the difficulties created by the American authorities regarding the sale of the Islander in the United States and highlighted the symbolic and psychological fallout of BN's collapse. A private company that had shown initiative and imagination – instead of waiting for public money – would disappear. In short, it would be the failure of a company that embodied the supposed remedies for the difficulties of the aerospace industry. But BN was a small company with little impact on the economic system, and therefore Prime Minister Edward Heath concluded that the government would not intervene to save it. The bankruptcy would run its course and, if anything, it might have been possible to find discreet ways to continue production of the Islander, given its importance for relations with Romania.[11]

The Romanians received news of the bankruptcy with surprising calm. Ambassador Pungan had talks with the FCO, BAC and the BN receiver, Maurice Eckman, appointed by Price Waterhouse. The lowest common denominator was the attention given to the continuation of the collaboration with IRMA, with which all parties involved were said to be more than satisfied. In particular, Pungan noted, the British were back to suggesting Romania's direct participation as a possible decisive contribution to a happy solution to the matter. So, it soon became necessary to begin discussions with Eckman and closely follow his negotiations with all BN's potential buyers in order to safeguard Bucharest's interests. The first contacts with Eckman aroused a cautious optimism and, on November 1, 1971, the Presidium approved the option of taking a direct investment in the future BN.[12]

The prospect of the participation of a Communist country in the British aviation industry actually stirred different, albeit not unexpected, reactions within the FCO. It was received favourably by those who followed the dossier of Anglo-Romanian bilateral relations, as is the case with Julian Bullard, Head of the East European and Soviet Department, while others showed a more cautious attitude. It should, however, be

---

11 UKNA, CAB 128/49/51: CM (71) 51st held on 21.10.71; UKNA, FCO 28/1532: Home to Bucharest, tel. 459, 21.10.71; "Receiver for Britten-Norman", *Flight International*, 28.10.71, p. 670.

12 ANR, CC, Chancellery, 1971_124: tel. 43.475, 27.10.71, ff. 31-34; protocol No. 31 al şedinţei Prezidiului Permanent al C.C. al P.C.R., 01.11.71, ff. 1-7; UKNA, FCO 28/1532: Laskey to FCO, tel. 353, 22.10.71; Home to Bucharest, tel. 474, 29.10.71.

noted that, although there were different points of view, no one seriously believed that the Romanian participation in BN would pose a threat to security. However, in view of the possibility of industrial espionage that the Romanian presence would have opened, there was the risk of being embarrassed by the possible reactions of the allies – specifically, that of the United States, which produced the engines of the aircraft, and could therefore object to their temporary transfer to a Warsaw Pact country. For the DTI, though, the problem was another one: had the Romanians enough money to make a credible bid?[13]

By comparing the information given by Eckman with that contained in a memorandum prepared by Cocoş, delivered illegally to the British by TAROM's press officer and deemed trustworthy by the FCO Security Department, the Romanians believed they were able to produce their own copy of the Islander using unpatented technical data – or, in any case, not registered in the United Kingdom – and the experience they had gained. Sales organisation, though, would have to stay in Bembridge, so that the aircraft could be sold more easily in Western countries. Therefore, even if the direct threat of industrial espionage seemed very minimal in this case, it is true that the Romanians were not willing to invest more than half a million pounds – a rather modest figure to acquire the whole of BN. The possibility of entering into partnership with the party that would take over the bankrupt company also remained open. At the end of 1971, the two most likely buyers seemed to be Fairey Engineering and Westland Aircraft. The latter, in particular, had a strong interest in the acquisition, mainly because it already owned BHC, who had worked as a subcontractor for BN, but the Romanians looked askance at such an eventuality, having reason to believe that Westland might decide to close down the Băneasa line.[14]

In January 1972, during the visit of Lord Drumalbyn, Minister without portfolio at the DTI in Bucharest, for the renewal of the 1967 technological cooperation agreement, Ispaş once more went back to the importance for Romania of participating in the commercialisation of the planes that would take place in the United Kingdom. More generally, however, the talks were essentially marked by rebukes to the British

13 UKNA, FCO 28/1532: minute by Lush, 16.11.71; Bullard to Lush, 16.11.71; Bottomley to Downey, 7.12.71. On the threat of Romanian intelligence, see Eliza Gheorghe, *The Romanian Intelligence Services During the Cold War: How Small Powers Can Sometimes Be Strong*, M.A. dissertation, Georgetown University, 2010.

14 UKNA, FCO 28/1532: note for the record, 11.11.71; notă cu privire stadiul acțiunilor întreprinse în legătură cu insolvabilitatea firmei Britten-Norman, 26.11.71; minute by Russell, 06.12.71; minute by Empson, 15.12.71.

government for its inaction during the BN crisis. Although the Romanians could accept that the possibilities for intervention were limited, they expected the British government to safeguard their interests, given its objective influence over the aviation industry. Verdeț noted that if fruitful cooperation could not be achieved, even in a restricted area such as that of the Islander, then Romania would have to reconsider the whole strategic choice of the United Kingdom as a partner for the reconstitution of its aviation industry. This option was motivated by the fact that London had the second biggest aircraft sector for expansion in the West, but concrete results had been discouraging, and the balance of trade between the two countries was now decidedly disadvantageous for Romania.[15]

By contrast, the relations between the Romanian authorities and the receiver were friendly and constructive. The problem was that BN's financial situation was not bad enough for Eckman to be in a hurry to liquidate the company. He admitted that the Romanian issue was important, also in political terms, but cash flows had resumed after the bankruptcy and he wished to take all the time necessary to find the best solution for settling past debts. Eckman's priorities, therefore, did not necessarily protect Romania's interest. Also, the Romanians were even more wary after realising that BN's conduct prior to the bankruptcy had deprived them of gains that they were entitled to. In fact, as per contractual agreement, the price of the Romanian labour component was linked to that of the materials provided by the company in Bembridge. If the latter increased, the former would have to do likewise for the benefit of the Romanians. But BN had preferred to provide materials to the Băneasa line at a loss in order to keep the final cost of the Islander unchanged, without, of course, informing its partners.[16]

In mid-February 1972, Pop launched a fresh initiative. He had also been hired as a consultant by Dowty-Rotol, a company specialising mainly in the manufacture of propellers, hydraulics and landing gear.[17] Dowty had every intention of snatching the contract from the French company Messier to supply landing gear, hydraulics and braking systems, as well as fuel control systems for YUROM, a Romanian-Yugoslav military plane that will be discussed in the next chapter. Pop's idea was that Dowty and Romania should buy BN jointly, with the Romanians acting as minority

---

15 UKNA, FCO 28/1978: record of meeting between Avram and Lord Drumalbyn, 05.01.72; record of meeting between Verdeț and Lord Drumalbyn, 10.01.72.

16 UKNA, FCO 14/1046: Hadley to Price, 12.01.72; Russell to Watts, 22.02.72; UKNA, FCO 28/2006: FCO to Bucharest, tel. 3/2, 10.01.72.

17 "Dowty To Acquire Rotol", *Flight International*, 26.12.58, p. 963.

shareholders, in exchange for the assurance that the YUROM contract be assigned to Dowty and not to the French. This solution would remove an obstacle that seemed to jeopardise the sale of other BAC aircraft to TAROM.[18] So, the production of the Islander at Băneasa became enmeshed in a tangle of relationships and transactions that went far beyond the intrinsic value of BN itself. As the British ambassador to Bucharest, Derick Ashe, commented:

> I believe that we are now at the crucial point in our relations with Romania in the aviation field. All the evidence suggests that we have an opportunity to tie up the whole package in a way, which, barring a complete reversal of Romanian policy, would secure our position for very many years to come. [...] The key to the situation is Dowty.[19]

Although Pop's idea was received very favourably by the FCO, it sparked a much less enthusiastic reaction from the Romanians. In fact, Ispaş and Tehnoimport noted that the various plans had to be kept separate and that each offer, allocated to a particular project, should make sense in itself and not simply in relation to the others. It should also be added that Dowty had no interest in the light aircraft field; it was willing to enter it not only because of the YUROM contract, but also because, with Romania's guarantee, a fair amount of materials for the Băneasa line would be imported. Even Sir George Dowty still had to be convinced of the validity of the operation and, in this regard, Ashe urged that the FCO should get the Secretary of State to explain the importance of this transaction for Anglo-Romanian relations. Finally, Dowty's reservations resulted in an offer for BN's acquisition well below Eckman's expectations.[20]

On this occasion, therefore, Pop's initiative was unsuccessful and, after several months of negotiations, BN was acquired by Fairey for 1.8 million pounds (more than three times the offer contemplated by the Romanians at the time). Fairey's purpose was merely to provide jobs at its plant in Gosselies, Belgium, by diversifying production in general aviation in the interval between the subcontracting work for the French Mirage and the work related to the F-16. Gosselies would, in other words, take on the role previously played by BHC, producing both the Islander and the

---

18 UKNA, FCO 14/1046: Ashe to DTI, CREDA. 13, 16.02.72; UKNA, FCO 28/2006: minute by Burns, 10.03.72; UKNA, FCO 69/294: Ashe to FCO, CREDA 16, 02.03.72.

19 UKNA, FCO 69/294: Ashe to FCO, CREDA 17, 02.03.72.

20 UKNA, FCO 14/1046: Burns to Russell, 24.03.72; note of a meeting between Onslow and Eckman on 10.05.72; UKNA, FCO 28/2006: Russell to Downey, 18.04.72.

Trislander, while the customisation and design activities would remain in Bembridge.[21]

Fairey had already had important relations with Romania when, in 1966, its nuclear branch had sold a small research reactor to Bucharest.[22] If the company remained interested in additional opportunities in this area, it would have no reason to close down the Băneasa line – relatively small, and yet producing according to excellent quality standards – whose materials would be, from that moment onwards, provided by Gosselies. It is significant, then, that BN's permanent representative in Romania, Mills, was promoted and transferred to the parent company with the task of following the integration of the production of the Islander and the Trislander in Gosselies. Fairey's CEO, Robert Holder, personally went to Bucharest after the acquisition of BN to decide with the Romanians on the timing for the completion of the 215 Islanders of the original 1968 contract and to offer the possibility of its continuation on the basis of market prospects. Not only did this reassure the Romanians about what they wanted most, it also served Holder as bait to try and lead Romania to invest in the Mainlander project, a short-haul passenger/cargo transport plane for 100 passengers or 10 tons of goods, on which a feasibility study was being completed. The aim was to pad out Gosselies, and the project was to be partly funded by the Belgian government; however, other partners were still needed to cover the costs of the launch and Romania could serve this purpose. Moreover, from the technical point of view, the production of the Mainlander would be fully within IRMA's reach (it was essentially a development of the Trislander, in turn, a development of the Islander) and the plane was one in which Bucharest had shown interest in the past.[23]

However, at this stage, the Romanians were already seeking to make a technological leap, especially in the aircraft engine sector, which could in no way be achieved with a plane of the same type as the one already in production at Băneasa and propelled by the old Dart turboprops produced by Rolls-Royce, with whom there had already been talks for a joint production of a jet engine, a topic that shall be addressed in the

---

21 "Fairey Buys Britten-Norman", *Flight International*, 07.09.72, p. 314; "Fairey Britten-Norman Plans", *Ibid.*, 28.12.72, p. 927.

22 Eliza Gheorghe, "Building Détente in Europe? East-West Trade and the Beginnings of Romania's Nuclear Programme, 1964-70", *European Review of History*, Vol. 21, No. 2 (2014), p. 235-253.

23 UKNA, FCO 14/1046: Russell to Hadley, 14.11.72; "And Now the Mainlander", *Flight International*, 14.12.72.

following chapters. The Mainlander project never saw the light of day, but Islander sales were booming (85 units were sold in the first six months following the acquisition by Fairey) and Britten declared publicly that the contract with IRMA would not be terminated; on the contrary, it could very well be extended. Production of the Islander, therefore, continued to be a very important element in Anglo-Romanian aeronautical relations, serving again for counter-trade in 1975, when BAC sold another six 1-11s to Romania to equip LAR (Liniile Aeriene Române), TAROM's charter company. In this way, a further 100 Islanders joined the original 215 that had been completed in October 1976.[24]

However, it should also be noted that in the mid-1970s, in a global context of recession and in the face of more and more obvious signs of the crisis of planned economies, the balance of ten years of relations was not being read positively. Now that the idea that Romania could become an important commercial outlet had been shelved, the country's importance tended to be based more on elusive political gains – affected, in turn, by the phasing out of Détente – and on occasional circumstances of economic profit. Inspired by the new contract for the Islander, the ambassador to Bucharest, Jeffrey Petersen, reflected:

> [T]here is at present some disenchantment with a connection, which, far from offering advantages to compensate for the sagging free enterprise markets, has shown itself equally vulnerable (and far less flexible) in its response to the exigencies of the recession. [...] At a time when pressures on us to get alongside the Eastern Europeans in all kinds of ways are slackening at home there is a parallel corrective move in the field, due to a more hard-headed appraisal of the problems and limitations of close collaboration with this country, and a more realistic assessment of the head of steam behind its economy.[25]

In this altered international context, the business strategy followed by Fairey proved suicidal. Faced with the worst recession of the postwar period, the Gosselies plant continued to churn out aircraft, at high costs now rendered untenable by the appreciation of the Belgian franc, with the aim of closing a few large contracts with government customers. The result was that the huge debt of the Belgian subsidiary would lead to the failure of the entire Fairey group in October 1977. Ominous signals had

24 ANR, CC, Foreign Relations, 1975_153: relațiile economice dîntre România și Marea Britanie, n.d., ff. 14-17; UKNA, FCO 28/2919: meeting in Bucharest of Lord Gorowny-Roberts, 20.10.76; "New Plans for Gosselies and Bembridge", *Flight International*, 19.04.73, p. 613.

25 UKNA, FCO 28/2919: Petersen to Cartledge, 06.10.76.

already appeared in 1976, when Britten and Norman, respectively Chief Executive and Sales Manager, left the company that still bore their name amid huge controversy over a marketing strategy that they did not approve of, as it ignored the qualitative characteristics of the aircraft, which actually accounted for its main strength[26]: '[n]obody has built a 'Land Rover of the Air': an aeroplane so simple and robust, so versatile in its potential applications, that ill-trained and unsophisticated operators could still use it successfully.'[27]

In May 1977, Ceaușescu authorised the opening of negotiations with Fairey to produce the Trislander in Romania. This would allow Băneasa to diversify its production with an aircraft that was certainly within IRMA's capabilities and that already had a good reputation on the world market, thus greatly reducing the costs for Fairey. By now, however, Fairey wanted to sell off the whole of BN to Shorts, a nationalised company under the control of the Northern Ireland Trade Department. The latter aimed to shift the entire production to Belfast to limit the redundancies (about 2000 on 5000 employees) at its Queen's Island plant, caused by the winding down of the activities of the Ministry of Defence. Following this operation, the Băneasa line would be closed down on termination of the existing contract.[28]

The Romanians were informed about the contacts with Shorts through their embassy in London only on August 4, 1977, less than two weeks before the publication of the final balance sheet for 1976 would lead to the suspension of Fairey quotations on the Stock Market, due to an excessive downward trend. Bucharest's reaction came a month later, when the Memorandum of Understanding with Shorts seemed imminent. The Deputy Minister of Foreign Trade, Bujor Almășan, sent word directly to Pop – thus avoiding diplomatic channels – that his government was available to set up a joint venture with Fairey, investing up to 5 million pounds. A letter from the Deputy Director General of Tehnoimport, Constantin Zgavîrdici, addressed to Holder, formalised the offer, in exchange for which the entire production of the Islander and the Trislander would be transferred to Romania. Fairey's CEO dashed off to Bucharest to meet the Minister of Foreign Trade, Ion Pațan, on September 10, four

26 UKNA, FCO 76/1543: minute by Baker, 27.09.77; UKNA, FV 14/90: note by Holder, 07.09.77; "Britten-Norman Move On", *Flight International*, 13.03.76, p. 624.

27 UKNA, FV14/90: market prospects for Islander aircraft, n.d.

28 ANR, CC, Chancellery, 1977_28: Mănescu la Ceaușescu, No. 27/XIII, 12.05.77; note H.2214, 14.05.77; UKNA, FCO 76/1543: FCO to Brussels, tel. 111, 18.08.77; "Shorts To Take Over Islander and Trislander", *Flight International*, 27.08.77, p. 574.

days before the planned signing of the Memorandum of Understanding with Shorts. It is worth noting that the British government was essentially a bystander during this animated phase, demonstrating the turnaround that bilateral relations between the two countries was experiencing. The Department of Industry (DOI) had no wish to back Fairey up, considering the company to be still solvent, whereas the Gosselies branch did not even have sufficient liquidity to pay wages in September. Instead the DOI had its sights on Shorts, which offered a purchase price deemed too low by the Belgians. At the same time, the workers in Gosselies prevented by force the removal of the machines that were supposed to be transferred to the United Kingdom. In this highly chaotic scenario, the possible effects on Anglo-Romanian political relations were not seriously taken into account.[29]

Caught between the Belgian imbroglio and the lack of money necessary to cover 51% of a possible partnership with Romania, the Fairey group was put under temporary receivership on October 11, 1977 and nationalised through the National Enterprise Board two weeks later, with the exception of the aviation component. The latter would be at the centre of complex trials and tribulations lasting several months, during which the priority in Whitehall was to prevent the British government from becoming involved in, or from appearing to favour, a particular solution to what was an essentially commercial issue. In a sort of déjà vu of the first bankruptcy, none of the initial solutions proved successful and, eventually, in 1978, BN was purchased by Pilatus, a subsidiary of the Swiss Oerlikon Group. Once again, as in 1975, this was possible thanks to the efforts of the former BAC – which, in 1977, was also nationalised and merged with HSA to become British Aerospace (BAe). BAe wanted to keep the Băneasa production line as an element of counter-trade in support of the most ambitious and controversial initiative in the history of the Anglo-Romanian aeronautical cooperation, the ROMBAC project. This meant that the Islander would continue to be produced in Romania until the end of this cooperation.[30]

29 UKNA, FCO 76/1543: Secondé to DOT, CREDA 74, 09.09.77; Secondé to DOT, CREDA, 75, 11.09.77; "Fairey Call for Help", *Flight International*, p. 1172-1173.

30 UKNA, PREM 16/1838: brief No. 14, n.d. UKNA, FCO 76/1543: Secondé to FCO, tel. 350, 14.10.77; FCO to Bucharest, tel. 228, 18.10.77; FCO to Brussels, tel. 182, 05.12.77; UKNA, DR 8/27: minute by Mawson, 14.02.80; "Pilatus To Buy Britten-Norman", *Flight International*, 05.08.78.

2.2 The Quest for Good Workmanship

In order to critically assess and evaluate the Anglo-Romanian aeronautical cooperation, it is necessary to consider its technical content. Romania's ambition to rebuild its aviation industry through relations with the United Kingdom has already been mentioned. On the one hand, there was an interest in the technology transfer and in industrial practices. On the other hand, considering the multiple pressures that the sector was subjected to, it was necessary to find new market outlets by offering relatively dated but not obsolete technology; this could very well live a second life in a different political and socio-economic context. However, the question arises as to whether and to what extent this combination worked or would work, since this is – after all – the dividing line between ambition and wishful thinking.

As has been stated, IRMA's initial task was to put together the planes starting from preassembled component kits that arrived from the United Kingdom. Help would be provided by British engineers – one from BN, resident in Romania, would be in charge of production, while an ARB engineer would be responsible for monitoring compliance with airworthiness standards. The good results achieved made it possible to loosen this strict supervision and, already by 1970, the Romanians were equipped with a set of jigs so that, from the thirteenth plane onwards, the Băneasa line independently produced components starting from raw materials imported from the United Kingdom. This cycle was interrupted by BN's first collapse, which resulted in the stoppage of the production until the Fairey takeover, a turning point for the Romanians who, from that moment onwards, began to work with relative autonomy on the wings of the Trislander.[31]

During the course of 1973, with the gradual return to normal of the Băneasa line, the inspector of the Civil Aviation Administration (CAA), Michael Inskip, was able to witness how labour standards were more than satisfactory and how, apart from a certain sloppiness in terms of keeping things clean and tidy, the Romanian staff were ready to act according to the surveys that had been carried out.[32] The level of organisational detail that the British engineers became involved with in Romania (recommending, for example, that the office of the Romanian Chief Inspector be located

[31] UKNA, DR 8/30: Guilbert to Owen, 26.08.70; UKNA, FCO 28/2349: survey: aviation in Romania, n.d.

[32] UKNA, DR 8/31: Romania Islander production line, report 2, 10-15.04.73; report 5, 04-07.12.73.

in a more silent area to allow him to do his job better or reporting the apparent poor state of health of some of the staff members) may seem surprising. This stemmed not so much from an inability to produce with high standards or to improve the technical quality of the work, but rather from an intrinsic inefficiency and sluggishness of the Communist system. So, long meetings where Inskip was essentially to play the role of mediator between the various branches of the apparatus were not at all unusual: '[t]hese discussions invariably help to sort out inter-ministry problems. One organisation will not accept instructions or suggestions from another organisation, but neither organisation will accept instructions and suggestions from a foreigner!'[33]

Even in the face of all the difficulties associated with the expansion of the assortment of products, the results appeared to be satisfactory. For example, there were delays in the production of the wings of the Trislander because the spars often came out of the assembly jigs, curved at the ends. After several unsuccessful attempts, the problem was identified in the excessive pressure applied during the squeeze riveting and it was finally solved. The difficulties, once again, were mainly organisational and systemic, such as training workers who were too young and inexperienced to comply formally with the organogram or the lack of cooperation between IRMA and the Romanian certification authorities. The situation became even more difficult when, in 1975, the consequences of management decisions made by the Fairey affiliate in Gosselies became evident, both in terms of shortage of supply and poor quality of the materials provided to the Romanian line from Belgium.[34]

BN's second collapse had far more relevant consequences on Băneasa, overlapping a gradual overstretch of the aviation industry in Romania, which will be discussed in the following chapters, and more generally to the deterioration of the economic situation in the country. Despite a general improvement in the situation in the early months of 1977 due to the presence in Romania of a representative of Gosselies, the interruption of supplies from Belgium following the bankruptcy meant that, in December, Inskip found a certain number of incomplete Islanders abandoned outside in the snow. This situation dragged painfully on during 1978, when 27 out of 31 Islanders were rejected by Inskip because they had been poorly assembled, their fuel circuit being contaminated with water and dust. As per usual, there was a lot of buck passing among the various Romanian

33 UKNA, DR 8/31: Romania Islander production line, report 9, 17-21.02.75.

34 UKNA, DR 8/31: Romania Islander production line, report 4, 18-23.09.73; report 7, 16-23.06.74; report 8, 1-5.10.74; report 11, 17-21.11.75.

authorities involved, but the energetic action of the CAA representative and of BN's Chief Inspector, Reginald Hobbs, ensured that things were back under control by January 1979.[35] Ten years after the beginning of the Islander production at Băneasa, in spite of two bankruptcies and a worsened context, this cooperation represented an undeniable success from the technical point of view and for the acquisition of skills by the Romanians. Indeed, as the CAA Chief Surveyor, Don Guilbert, noted, surely with a hint of annoyance, 'it is in fact somewhat regrettably true that the Romanian standards of manufacture are generally somewhat better than the standards normally achieved in the United Kingdom.'[36]

The collapse of the Fairey group also brought back to the fore the issue of flight certification for the sale of the Romanian-produced Islander in the United States. In 1976, a bilateral agreement had been reached between Washington and Bucharest, but it only concerned gliders. In the summer of 1977, the British ruled out changes to the procedures. The 250-strong units produced from 1969 onwards continued to receive certification by the CAA to be registered in the United Kingdom or exported to non-FAA countries.[37] Moreover, the marketing of the plane through Bembridge had also been, right from the beginning, a condition for avoiding problems with the Allies in CoCom and it had then become a form of protection for BN, as a means of maintaining control over sales policy. As Dennis Berryman, BN's technical director while the company was under temporary receivership, clearly stated, 'it is not our wish that IRMA should ever be in a position to supply aircraft direct from Romania to the United States, thus we build in the safeguard that all aircraft are only exported thought the Bembridge system.'[38]

The freeze in Gosselies and the controversy over the fate of the Belgian plant profoundly changed the terms of the problem and the possibility of selling the Romanian Islanders in the United States became a pressing matter. Now BN was willing to draw up an export certificate for the United States, provided that the FAA accepted it. In this regard, Lawrence Edwards, formerly a CAA Safety for Air Attaché in Washington and now employed as a consultant by BN to solve the problem, contacted the head of the FAA Aircraft Certification Department, Don Jacobsen, to

---

35 UKNA, DR 8/31: Romania Islander production line, report 4, 18-23.09.73; report 7, 16-23.06.74; report 8, 1-5.10.74; report 11, 17-21.11.75.

36 UKNA, DR 8/32: Guilbert to White, 12.10.79.

37 UKNA, DR 8/27: meeting between BAC and CAA, 08.07.77; UKNA, DR 8/32: Edwards to Guilbert, 19.09.77.

38 UKNA, DR 8/32: Berryman to Guilbert, 30.11.77.

test the waters. Edwards upheld the high standard of workmanship of the Băneasa line, explaining in detail the procedures of quality control that were applied upon agreement with the CAA and noting that 42% of the content of the plane had originated in the United States, starting with the Lycoming engines. Considering all this, would the FAA accept the Romanian Islander when offered? Jacobsen said that, actually, there was no written policy on the matter, because one had never been formulated. Since the CAA insisted that the Romanian planes were, in every aspect, identical to, if not better than, those produced in Belgium and in the United Kingdom, in terms of safety there were no problems. However, an economic and political assessment of the issue was not ultimately down to his department.[39]

In fact, it had long been known that the problem was not a technical one, but the fact that the FAA continued to maintain that the Islander produced at Băneasa could not be accepted because it was not covered by the bilateral agreement with Romania, nor by that with the United Kingdom. The last Gosselies batch was rapidly running out, and if this situation did not change, it would mean serious trouble for BN, especially when the process of acquisition by Pilatus was nearing completion. Thus, the pressures on the Americans made by the company and by the CAA rapidly intensified. At the end of 1979, the FAA agreed to allow imports of Romanian Islanders, provided that the CAA guaranteed the entire production cycle under its sole responsibility. Contrary to what had happened in the past, now all the machines would be completed in Bembridge and the CAA could assume the same level of responsibility for all the Islanders, regardless of where they had been produced. This is proof of how the technical quality of the product offered by the Romanians had never really been in question.[40]

39 UKNA, DR 8/32: Edwards to Jacobsen, 14.02.79; memorandum by Jacobsen, 13.03.79.

40 UKNA, DR 8/32: Berryman to Guilbert, 14.11.79; Plummer to Chaplain, n.d.

Chapter 3

Military Aviation: JUROM and Beyond

In 1987, in the well-known book *Red Horizons*, the former Acting Chief of the DIE (Departamentul de Informații Externe, Department of External Intelligence) and Ceaușescu's close collaborator, Ion Mihai Pacepa, who had enjoyed political asylum in the United States for nearly a decade, revealed how, at a meeting in Brioni in 1973, Tito had proposed to the Romanian leader a secret partnership between the two countries for the development of a military aircraft to be exported to the Third World. The JUROM project, a name made up by the initials of the two countries, was supposed to benefit from Western aid thanks to the attitude maintained by Belgrade and Bucharest towards Moscow.[1]

Pacepa's account is inaccurate and deliberately misleading, although it is true that the JUROM represented one of the most important elements of the Romanian-Yugoslav cooperation of that period.[2] The project had originated years earlier and it formed part of the Romanian programme of recovery of its aviation industry, as devised in the late 1960s. As such, the JUROM project shows not only the relevance of Yugoslavia in the eyes of Ceaușescu's Romania, but it also turns out to be the origin of a network of military aviation relations between London and Bucharest, which allow to gauge both the degree of impatience reached in the 1970s towards the constraints imposed on East-West trade, and – at the same time – the lasting effects of the Brezhnev Doctrine and of the Romanian alliance with the Soviet Union.

3.1 The Brezhnev Doctrine as a Propeller

On the night between August 20 and 21, 1968, the Soviet Union, along with other Warsaw Pact countries, invaded Czechoslovakia, reaffirming unequivocally to what extent it was acceptable to question the logic of the

1 Mihai Pacepa, *Red Horizons. Chronicles of a Communist Spy Chief*, Washington, Gateway, 1987, Italian translation, Trento, L'Editore, 1991, p. 326-327.

2 ANR, CC, Foreign Relations, 1974_139: probleme actuale ale relațiilor româno-jugoslave, undated note by Pungan, ff. 79-84.

blocs. Since January 1968, top Soviet politicians had watched the parabola of the Czechoslovak Communist Party with growing apprehension and they feared that the country, which was fundamental to the geopolitical and military architecture of the Warsaw Pact, could follow in the footsteps of Yugoslavia or Romania. In addition to the fear of the possible counter-revolutionary effects of the Prague Spring, there was the threat of a possible Balkan agreement between Czechoslovakia, Yugoslavia and Romania, which would be an open challenge to Soviet hegemony and a serious risk to the survival the Warsaw Pact.[3]

Romania was not one of the member countries that took part in the invasion and, not surprisingly, it had been side-lined from the decision-making process, which resulted in the military operation of August 1968, nor had it been consulted. During the previous year, the challenges deriving from Bucharest's foreign policy had multiplied. These included firstly the recognition of the FRG, although it was understood that the Warsaw Pact countries would not normalise relations with Bonn until the latter accepted the Oder-Neisse Line; secondly, the fact that Ceauşescu had not participated in the summit of the first secretaries of the various Communist parties on the situation in the Middle East held in Budapest in July 1967; thirdly, the 'major departure'[4] from the discipline of the bloc, namely the continuance of diplomatic relations with Israel after the Six-Day War; fourthly, the over resistance to Moscow's attempts to integrate the Eastern bloc; lastly, equidistance in the Sino-Soviet split. In 1968, Romania's open support to Czechoslovakia, culminating in Ceauşescu's visit to Prague and in the treaty of friendship between the two countries, which had been agreed upon just before the Soviet invasion, further reignited Moscow's fears. Therefore, Romania's exclusion from the discussions on Czechoslovakia's fate not only fitted with the Soviet need to reach prompt decisions but was also an implicit warning to Bucharest.[5]

---

3 Kieran Williams, "The Russian View(s) of the Prague Spring", *Cold War History*, Vol. 14, No. 2 (2012), p. 128-141; Mark Kramer, "The Prague Spring and the Soviet Invasion in Historical Perspective", in Günter Bischof, Stefan Karner and Peter Ruggenthaler (eds.), *The Prague Spring and the Warsaw Pact Invasion of Czechoslovakia in 1968*, Plymouth, Lexington Books, 2010, p. 33-58.

4 Fischer, *Nicolae Ceauşescu*, *op. cit.*, p. 98.

5 Kieran Williams, *Romania's Resistance to the USSR*, in Kristen Williams, Steven Lobell and Neal Jesse (eds.), *Beyond Great Powers and Hegemons. Why Secondary States Support, Follow, or Challenge*, Stanford, Stanford University Press, 2012, p. 33-41; Mircea Munteanu, "When the Levee Breaks. The Impact of the Sino-Soviet Split and the Invasion of Czechoslovakia on Romanian-Soviet Relations, 1967-1970", *Journal of Cold War Studies*, Vol. 12, No. 1 (2010), p. 43-61; Mark Kramer, "The Czechoslovak Crisis and the Brezhnev Doctrine", in Carole Fink, Philipp Gassert and Detlef Junker

The exclusion was itself enough to justify the surprise and the alarm of the Romanian leadership when Czechoslovakia was invaded, and it also explains their ensuing attempt to follow a middle course between the anti-hegemonism embodied in the principles of the Declaration of April 1964, as well as a greater willingness to compromise with the Soviets.[6] In any case, the first reaction was one of open condemnation and defiance, with the speech delivered by Ceauşescu from the balcony of the headquarters of the Central Committee.

> Penetration of the troops of the five socialist countries into Czechoslovakia is a big mistake, a grave danger for peace in Europe, for socialism in the world. It is inconceivable that in today's world, when people are rising up to fight in defence of national independence and for equal rights, socialist states are trampling on the freedom and independence of another state. There is no justification, not even for a moment can we accept the idea of military intervention in the internal affairs of a socialist sister country. (*Applause, loud and prolonged cheers*). […] They say that in Czechoslovakia there is a counter-revolutionary threat: maybe tomorrow some people will say that even here, at this meeting, counter-revolutionary tendencies are present. To all of them we say: the entire Romanian people will not allow anyone to violate the territory of their country. (*Stormy applause, long cheers*).[7]

Actually, the historical evaluation of what was probably the Romanian dictator's finest hour is quite controversial. Tismăneanu calls it a masquerade, a manipulation of the anti-Soviet card in order to obtain a genuine internal political consensus and to be perceived internationally as a maverick, whereas in fact he was building up an ultra-authoritarian, personal dictatorship based on the motto *Partidul, Ceauşescu, România*.[8] Dennis Deletant, while noting how the appeal to national sentiment was becoming an effective instrument of social control, emphasises that Ceauşescu's speech in any case represented a courageous act in the face of the real fear that Romania might, in turn, be invaded.[9] Mihai Retegan and Cezar Stanciu, though in different ways, stress how Ceauşescu's attitude shifted very quickly from confrontational to appeasing towards

(eds.), *1968: The World Transformed*, Cambridge, Cambridge University Press, 1998, p. 136-137.

6 Mihai Ionescu, "Introduction", in *Romania and the Warsaw* Pact, *op. cit.*, p. 79-83.

7 *Cuvîntul tov. N. Ceauşescu la adunarea populaţiei din capitală în piaţa palatului republicii*, in *Principiile de bază ale politicii externe a României*, Bucharest, Ed. Politică, 1968, p. 11, 13.

8 Tismăneanu, *Stalinism for All Seasons*, *op. cit.*, p. 202-203.

9 Deletant, 'Taunting the Bear', *op. cit.*, p. 499-500.

the Soviets, after both the Yugoslavs and the Chinese had made it clear that they were not able to help Romania concretely in the event of invasion. Therefore, the best thing for Bucharest would be to seek reconciliation with Moscow, while trying to save the principles of its foreign policy as far as possible.[10] Mircea Munteanu draws attention to how, on the one hand, manipulating the condemnation of the invasion of Czechoslovakia in order to win the support of the West and, on the other, adopting a softer policy toward Moscow to normalise the position of Romania within the Warsaw Pact run parallel and are not mutually exclusive. Consequently, Romania's foreign policy would show a consistent direction, with Ceauşescu's speech crowning a decade of growing impatience with the Soviet hegemony. Not even the compromises with Moscow would affect the bases of this policy.[11]

This debate reflects, perhaps excessively, a strong historiographical connotation of the term 'appeasement,' thus overshadowing the fact that even before the invasion of Czechoslovakia, nationalism and anti-Sovietism were used to strengthen the legitimacy of the regime and to expand the area of manoeuvre in foreign policy. At the same time, the opening to the West certainly did not mean the interruption of relations with the USSR, but rather an attempt to play off the parties involved against each other in order to get as many benefits as possible from both. Of course, talking about a fundamental continuity is not to say that it was simply play-acting by the Romanian regime nor to deny the importance of the measures taken or that of the tactical adjustments already established. Furthermore, Mănescu himself had already suggested that the meeting with Tito on August 24, 1968 had a crucial role in convincing Ceauşescu of the opportunity to follow a more prudent course of action with the Soviets, and that concerns that Romania could be made the target of a new invasion, aimed at eliminating its leadership, were certainly real.[12]

While it is true that, along with the Brezhnev Doctrine, the long-discussed reform of the Warsaw Pact came into effect, strengthening the Soviets' supervisory power over the member countries, it is also true that Ceauşescu opposed the new Statute of the Joint Armed Forces, which would entail supranational integration at the expense of the national communist parties. The Romanian participation in the Brotherhood-in-Arms, the first

---

10 Mihai Retegan, *In the Shadow of Prague Spring: Romania Foreign Policy and the Crisis in Czechoslovakia, 1968*, Iaşi, Centre for Romanian Studies, 2000; Cezar Stanciu, "Crisis Management in the Communist Bloc: Romania's Policy Towards the USSR in the Aftermath of the Prague Spring", *Cold War History*, Vol. 13, No. 3 (2013), p. 353-372.

11 Munteanu, "When the Levee Breaks", *op. cit.*, p. 43-44, 56.

12 Mănescu, *Convorbiri nedeterminate*, *op. cit.*, p. 202-205.

military exercise involving the Pact countries in 1970, was limited to a few observers. The new Romanian-Soviet bilateral treaty, unlike the one between Moscow and Prague, did not envisage mutual defence obligations except against external threats and it made no reference to the defence of socialist gains. The 1972 Law on National Defence excluded – at least juridically speaking – a repeat of what had happened in Czechoslovakia, stating that only the elected Party and state bodies could request external assistance.[13] At the end of 1978, when the Soviets demanded greater military preparedness and an almost total interoperability of the main categories of weapons, justifying such measures with the threat posed by the modernisation of NATO forces, Romania still opposed the analysis of the situation and asked for a reduction of the troops and military expenses of the Warsaw Pact.[14] Similarly, in the CMEA, even with less room to manoeuvre than in the past, Bucharest would continue to act as a brake on discussions about economic integration, especially because the government did not want their relations with the EEC to be compromised, at a time when the latter was in the process of applying a common economic policy.[15]

Romanian caution is undoubtedly evident in these bilateral relations. When in 1969 the Soviets decided to end the system of low interest loans which regulated Romanian imports of military technology, the Permanent Praesidium gave clear instructions to follow a soft approach, endeavouring to negotiate a different payment solution and simultaneously planning the extinction of the accumulated debt.[16] However, although the USSR's leverage on Romanian foreign trade declined during the 1970s, special attention continued to be paid to technological cooperation, especially in the military field, for example with the production in Romania of the amphibious vehicle for transporting troops, the BTR-70, and the YAK-52 trainer aircraft.[17]

---

13 Matthew Ouimet, *The Rise and Fall of the Brezhnev Doctrine in Soviet Foreign Policy*, Chapel Hill, University of North Carolina Press, 2003, p. 72-83.

14 Opriş, *Industria românească de apărare*, *op. cit.*, doc. 68, p. 321-334.

15 Suvi Kansikas, *Socialist Countries Face the European Community. Soviet Bloc Controversies Over East-West Trade*, Brussels, P.I.E. Peter Lang, 2014, p. 93-104, 117-130, 160-163.

16 ANR, CC, Chancellery, 1969_113: protocol No. 29 al şedinţei PP din ziua de 18.08.69, ff. 2-5; notă 180/848, 10.07.69, ff. 21-25.

17 "Notă referitoare la stadiul relaţiilor bilaterale româno-sovietice", May 1977, in *România şi Tratatul de la Varşovia. Conferinţele miniştrilor Afacerilor Externe şi ale adjuncţilor lor (1966-1991)*, Documente Diplomatice Române, Seria a III-a, Bucharest, Alpha, 2009, p. 521-528.

Thus, the Romanian attitude remained basically unchanged. In the famous May 1970 meeting with Brezhnev, during which the Secretary General of the CPSU listed a long string of Soviet grievances, along with some ill-concealed threats, Ceauşescu made the following statement, with reference to relations with the West:

> We said that we would be satisfied to import from socialist countries, we even handed them a list [of goods], but they told us that some of those are not available. Then, of course, we will take from where we can. In any case, we have decided to ensure the future development of the Romanian economy, and if we cannot resolve it with the socialist countries, we will resolve it through any means available to us.[18]

In this context, however, the invasion of Czechoslovakia had specific and significant repercussions on Romania's aeronautical ambitions and, consequently, on relations with the United Kingdom in this sector.

In February 1968, immediately after signing the contract with BAC for the purchase of 1-11s, a note to the Executive Committee of the RCP made it clear that the recovery of the aviation industry would entail the construction of a fighter jet, a medium-small passenger plane, a trainer and a utility aircraft. In the last three cases, the structure would be developed independently, while the engines would be built under licence. Even more remarkably, priority was given to supersonic fighters, taking into consideration both the French Mirage F3 and the Lightning F Mk-3, developed by English Electric and later taken over by BAC, in view of the lack of adequate knowledge on the latest Soviet developments in the field.[19] At the Executive Committee meeting on February 27, Maurer sharply criticised this approach and he underlined the senselessness of proceeding with the reconstitution of an independent capacity for aircraft production starting from the most technically complex element, that is, the fighter jet. During the long debate, Ceauşescu intervened several times to settle the criticisms: the plane was to be built with the cooperation of a partner – be it the USSR, France or the United Kingdom – and Maurer's concerns were, therefore, unjustified. Moreover, Ceauşescu added, it was necessary for Romania to be able to produce most of the equipment needed in case of war by itself; from this point of view, the navy and the air force were the most vulnerable.[20]

---

18 Mastny – Byrne (eds.), *A Cardboard Castle?*, *op. cit.*, p. 374.

19 ANR, CC, Chancellery, 1968_29: notă cu privire la studiile preliminare privind stabilirea Industria Aeronautică Română, n.d., ff. 38-59.

20 Opriş, *Industria românească de apărare*, *op. cit.*, doc. 48, p. 244-250.

Two points should be raised. First, even in the military sector, we find the same approach described in previous chapters regarding the modernisation of the TAROM fleet. The possible choice of a Western partner was not in itself anti-Soviet, but rather it was believed that – once a certain goal had been fixed – Romania was free to accept the best offer, perhaps playing off the bidders against each other. Thus, the interest shown in the French and British fighters did not prevent Ceauşescu from telling the Soviet ambassador, in the fateful spring of 1968, that the USSR must grant licence for the production of aeronautical material in Romania. This would not only meet Romania's needs, but also make a profit; however, the Secretary General declared, "trade is trade, it does not only depend on the buyer, but also on the seller; if he who sells does not want to sell, buyers cannot buy".[21] Secondly, it was not only a question of boldness towards the supplier, something consistent with the principle of an anti-hegemonic foreign policy. Indeed, in the months preceding the invasion of Czechoslovakia, striking terms were used to express the desire to have an autonomous military production in case of crisis or war. In addition, besides reflecting a general aspiration to industrial development, such terms could not fail to recall the example of Yugoslavia – a country to which Romania was now linked by a relationship of friendship and close cooperation.[22]

The similarity of the paths followed by Yugoslavia and Romania is well known. By the end of the 1970s, Romania was showing an interaction pattern at global level, which was very similar to Yugoslavia's, with a significant increase in interchanges with the less-developed countries and a consolidated partnership with the capitalist West in terms of investments and credits.[23] Although bilateral trade stood at a low level, and was deemed unsatisfactory,[24] nevertheless Romania could see in neighbouring Yugoslavia a profitable example of how to get the greatest possible benefit from the new opportunities offered by distension and decolonisation. At the same time, the robustness of Romanian economic and industrial development in this period aroused interest and admiration in Yugoslavia,

---

21 ANR, CC, Foreign Relations, 1968_69: notă de audienţă, 22.05.68, f. 3.

22 Laura Herta, *De la relaţiile româno-sârbe la relaţiile româno-iugoslave. Interpretări constructiviste*, Cluj, Ed. Fund. Studii Europee, 2012.

23 Ronald Linden, *Communist States and International Change. Romania and Yugoslavia in Comparative Perspective*, Boston, Allen & Unwin, 1987, p. 53-54, 68, 79-80.

24 ANR, CC, Foreign Relations, 1968_76: propuneri privind problemele care ar putea constitui obiectul convorbirilor cu prilejul vizitei în RSF Iugoslavia, 10.06.68, ff. 58-65.

especially in terms of technical and instrumental equipment.[25] Thus emboldened, Ceauşescu was able to take the initiative to propose a far-reaching cooperation to the Belgrade government, both in obtaining raw materials necessary for industrial development, and in securing commercial outlets – with particular attention to the Chinese market. It was a differentiation process intended to explicitly reduce the relevance of the USSR and the CMEA, with cooperation being the magic formula to solve problems of development and security.[26]

Although it is impossible to say definitely whether Romania's aeronautical ambitions were inspired by the example of Yugoslavia, the similarities cannot be ignored. In the context of the creation and setting-up of the SOKO aircraft factory in Mostar, in the summer of 1951, the Yugoslav government began to urge the United Kingdom to license the manufacturing of Goblin and Vampire jet fighters, arguing that they wanted to revive the pre-war aviation industry employing British know-how. Obviously, the Yugoslav trump card was the fact of being on bad terms with the Stalinist USSR and therefore, according to a binary logic typical of the Cold War, worthy of support from the West. Tito raised the question directly with Eden in September 1952, highlighting the weak position that Yugoslavia would find itself in when Moscow equipped its Balkan satellites with modern aircraft.[27] In Whitehall, the debate focused primarily on financial solvency and on Yugoslavia's industrial capacity, rather than on military-security issues. Everything was further complicated by the American variable in the context of the multilateral export control mechanism and the obvious attempt by Belgrade to play not only on the bipolar conflict, but also on the commercial competition between Western countries.[28]

From the end of 1955, Yugoslavia showed interest in the Gnat, a subsonic jet trainer and light fighter aircraft that Folland was developing. The structural characteristics of the plane made it ideal, both in terms of use and because there was no need for specialised tools at production level. Therefore, it could also be constructed by a relatively backward industry

25 ANR, CC, Foreign Relations, 1970_6: stenogramă discuţiilor tov. N. Ceauşescu cu delegaţia Consiliului Central al Uniunii Sindicatelor din Iugoslavia, 30.03.70, ff. 3-14; notă cu privire la vizita delegaţiei Uniunii Sindicatelor din Iugoslavia, 27.03.70, ff. 15-24.

26 ANR, CC, Foreign Relations, 1970_6: stenogramă discuţiilor tov. N. Ceauşescu cu preşedintele părţii iugoslave din Comisia mixtă româno-iugoslavă, 17.09.70, ff. 2-10.

27 UKNA, FO 371/95555: Quincey to Pearson, 23.08.51; minute by Davies, 31.08.51; UKNA, FO 371/102241: record of meeting with Marshall Tito on 22.09.52.

28 UKNA, 107875: minute by Cheetam, 8.07.53; UKNA, FO 371/118057: Belgrade to FO, tel. 572, 10.07.55.

such as that of Yugoslavia. The United Kingdom thus began to see the sale of the manufacturing licence for the Gnat as a profitable opportunity to establish a firm link to the development of Yugoslavian aviation, showing more and more availability in relation to the risk that Belgrade might turn to USSR for modernisation of its air industry.[29] The reluctance of the Conservative government in London to finance the transaction with export credit and the limited room for manoeuvre of Folland, a relatively small company, soon to be absorbed by Hawker-Siddeley, certainly did not favour the already lengthy and trouble-fraught negotiations. Yugoslavia acquired only two Gnats in the summer of 1958 for evaluation purposes, while resources required for the acquisition of the licence were used instead to buy seventy-eight American F.86Es, which would have obviated the immediate operational needs.[30]

Perhaps inspired by the promising prospects of the Aermacchi MB-326 in nearby Italy, in 1957, the Belgrade government decided to independently design a robust and simple trainer aircraft, characterised by an overall economy of structure, which would fulfil the role of the Gnat. This was the Galeb, which made its maiden flight in May 1961. The following year its single-seater attack version, the Jabstreb, was developed. Its structure was substantially identical, but with more sophisticated electronic and navigation systems. Both planes were presented at the 1964 Farnborough Air Show, powered by different versions of the Armstrong-Siddeley Viper engine.[31]

Yugoslavia's relations with Armstrong-Siddeley dated back to 1928, when the then Bristol-Siddeley sold the Industria Aeroplanskih Motora of Rakovica the licence to produce the air-cooled radial Jupiter, the first aircraft engine manufactured in what was still the Kingdom of Serbs, Croats and Slovenes. Immediately after the Second World War, Communist Yugoslavia had shown renewed interest in British engines, prompting a positive reaction from the British government, especially from Prime Minister Clement Attlee, but the support given to the Greek Communists during the civil war made it politically impossible to follow this up.[32] In 1952, a new application for licence to produce turbojet engines

29 UKNA, FO 371/118058: extract from AWP (55) 29 held on 08.11.55; UKNA, FO 371/130549: extract from AWP (57) 5 held on 02.04.57; Nicholls to Ross, 13.12.57.

30 UKNA, FO 371/130549: Young to Nicholls, 24.06.57; UKNA, FO 371/145208: Garvey to Addis, 12.08.59; minute by Galbraith, 11.09.59; minute by Ballantyne, 08.10.59.

31 "SOKO Galeb and Jabstreb", *Flight International*, 28.11.68, p. 901-905.

32 UKNA, FO 371/59541: Belgrade to FO, tel. 246, 18.02.46; signed letter to the Under-Secretary of State for Air, Air Ministry, 28.02.46; UKNA, FO 371/72596: Wallinger to the Secretary to the Board of Trade, 29.01.48.

was rejected at the last moment because of concerns on the part of the Americans, who apparently feared the strategic consequences of the transfer of technological know-how related to modern axial-flow compressors, which were much more efficient than the centrifugal type.[33] In 1955, the export of ten Viper engines to Yugoslavia was finally authorised. These were equipped with less powerful axial compressors than older models. So, over the next decade, Armstrong-Siddeley (taken over by Rolls-Royce in 1966) would have a turnover of millions of pounds, by providing Viper engines to power both the Galeb and the Jabstreb, respectively in the 11.1 kN Mk. 22/6 version (the same engine as the Aermacchi MB-326) and in the 13.8 kN MK. 531 one (that of the MB-326G and the BAC Strikemaster).[34]

Therefore, Yugoslavia can be said to set a concrete example for Romania to follow in a specific area. This embodied, firstly, the very idea of a revamp of the national aviation industry, combining the need for prestige with more objective requirements determined by Romania's peculiar geopolitical position; secondly, its implementation through both local effort and the partnership with the United Kingdom, which involved the import of components and technology; thirdly, the systematic use of tactics to oppose capitalist countries and the Soviet Union, while furthering commercial competition in Western Europe; fourthly, the fitful and erratic progress of negotiations. Even more specifically, albeit within a programme so ambitious as to seem tinged with megalomania, one of Romania's aims was to design and produce a trainer, precisely along the lines of the Galeb and the Jabstreb, with the important difference that the engines would not be bought, but produced under licence. Significantly, the Technical Director of Rolls-Royce, Stanley Hooker, recalls how in 1965 – that is, before the Bârlădeanu's visit to the United Kingdom – a Romanian delegation had arrived in Bristol, led by Ispaş, who had shown great interest in the Viper.[35]

The invasion of Czechoslovakia would finally cause Romanian and Yugoslav paths to converge with regard to aeronautical development and would result in a sharp acceleration of events. At the same time, it also

---

33 UKNA, FO 371/102242: minute by Harrison, 19.11.52; minute by Harrison, 28.11.52.

34 UKNA, FO 371/107875: Orchard to Leavett, 04.11.55; UKNA, FO 371/118057: Leavett to Tupman, 15.06.55; extract from AWP (55) 1 meeting held on 07.06.55; "Industry International", *Flight International*, 21.11.63, p. 850.

35 Hooker, *Not Much of an Engineer*, *op. cit.*, p. 199.

meant that the United Kingdom's cooperative relationship with a Warsaw Pact country would prove inconvenient and politically sensitive.

In fact, among the measures taken by the regime immediately after the Prague Spring, such as the reinforcement of the work of the Council for State Security's information network,[36] and, in October 1968, there was the activation of the Institute for Aerospace Design and Construction (ICPAS, Institutul de Cercetări și Proiectări Aerospațiale) and the Headquarters of the Romanian Aeronautical Industry (CIAR, Centrala Industriei Aeronautice Române), responsible for R&D and manufacturing respectively. The aim was to design and produce a military jet in three-to-four years. The Institute of Fluid Mechanics of the Academy of Romania, headed by the renowned scientist Elie Carafoli, would be merged with ICPAS to contribute to help design a training and ground attack aircraft in cooperation with Yugoslavia, while the CIAR, being an organisation of the Ministry of Machine Building, would take care of production activities and supply materials. Thus, despite the possibility of official denial or dissimulation, there was a massive conversion to practical and military purposes of resources previously dedicated to fundamental research, as in the case of the Institute of Fluid Mechanics.[37]

At the same time, attempts were made to get the British and the French to cooperate on a supersonic fighter, as determined by the Executive Committee. With regard to the British, among the conditions to proceed with the purchase of BAC 1-11s, in February 1968, the Romanians had sought Rolls-Royce's cooperation to produce an engine with a thrust within the 66.6 to 88.8 kN range. For what concerns the French, during Charles De Gaulle's sensational visit to Bucharest in May 1968, Ceaușescu and Maurer had asked France's help to organise the production of military aircraft in Romania. The French president, though surprised, had suggested considering not only the aircraft, but also the engines.[38] Thus, between September 26 and October 6, 1968, a joint technical mission of the Ministry of Machine Building and the Ministry of Foreign Trade was sent to Britain and France in the strictest confidence – indeed not even

---

36 Elis Neagoe-Pleșa, „1968 – anul reformării agenturii Securității", *Caietele CNSAS*, Vol. 1, No. 1 (2008), p. 9-22.

37 National Archives and Records Administration (NARA), Access to Archival Databases (ADD), RG 59 Central Foreign Policy Files, Bucharest to SecState, tel. 1738, 10.05.73; Mihai Neamțu, "Remembering Professor Elie Carafoli", *INCAS Bulletin*, No. 2 (2009), p. 42-48.

38 UKNA, FCO 28/247: note of a meeting by Meynell, 21.02.68; Historical Archives of the European Communities (HAEC), documents of the French Ministry of Foreign Affairs (MAEF), reel 34, entretien franco-roumain tenu a Bucarest le 15.05.68.

the Romanian ambassador in Paris had been forewarned[39] – to discuss the terms of the sale of licences and the necessary technical assistance.

The Romanian expectations, however, were to be disappointed because of the repercussions of the events in Prague. Operation Danube had taken the British government by surprise. Wilson and Stewart returned from holiday for an emergency Cabinet meeting on August 22, but aside from public condemnation and some concern that Romania and Yugoslavia could meet the same fate as Czechoslovakia, extreme caution was taken not to jeopardise the Détente process or to put economic relations with Eastern Europe at risk, due to initiatives that, in any case, would not resurrect the Prague Spring. This was accompanied by a renewed urgency to maintain cohesion between the Western countries and the fear that the disillusionment generated by the invasion of Czechoslovakia would stimulate the interest of the USSR satellites against in the 'free world.'[40]

Before the FCO could send instructions to the diplomatic posts in the Warsaw Pact countries to return to business as usual, the Romanian delegation arrived in the United Kingdom. Here, by order of the government, Romanian delegates were prevented from visiting BAC's Preston division, which produced military equipment, while Rolls-Royce informed them that, in the new situation, negotiations to obtain the licence to manufacture the 25R Spey engine – launched in Bucharest in the weeks before the events in Prague – would have to be interrupted. The Spey was a turbofan engine developed in the early 1960s and it was used both on the HSA Trident, and on the BAC 1-11. Military variants had also been produced, among them the RB.168-25R. This model, which was more ambitious, was fitted with an afterburner developed thanks to the experience gained with the engine of the Lightning fighter and it had a thrust of 93.2 kN. Rolls-Royce, in fact, had signed a fixed price contract, with guarantees of costs and performance which would allow the Spey to be used on the F.4M Phantom fighters instead of the General Electric J79 turbojet,

39 HAEC, MAEF-34: entretien avec l'ambassadeur de Roumanie du 25.09.68.

40 Gil Bennett and Keith Hamilton (eds.), *Britain and the Soviet Union, 1968-72*, doc. 23, in *Documents on British Foreign Policy Overseas* (DBPO), Series III, Vol. 1, London, HMSO, 1997, p. 111-115; Saki Dockrill, "Defence and Détente: Britain, the Soviet Union, and the 1968 Czech Crisis", in Bischof – Karner – Ruggenthaler (eds.), *The Prague Spring, op. cit.*, p. 249-270; Geraint Hughes, "British Policy Towards Eastern Europe and the Impact of the Prague Spring: 1964-68", *Cold War History*, Vol. 4, No. 2 (2004), p. 115-139; Anne Deighton, "*Ostpolitik* or *Westpolitik*? British Foreign Policy, 1968-75", *International Affairs*, Vol. 74, No. 4 (1998), p. 893-901.

despite the higher initial production cost of a turbofan as compared to a turbojet.[41] Due to the problems of combustion instability in the afterburner which delayed the implementation of the programme, Rolls-Royce stood to benefit from the Romanian interest, but – as Benn wrote in his diary – after the events in Prague, Romania was no longer seen as a kind of Finland,[42] a country which could usefully receive the assistance of the West, but primarily as a Warsaw Pact country, even though Ceaușescu had condemned the invasion.[43] What Rolls-Royce could negotiate at the time was the construction of a workshop for assembly and repairs of the civil version of the Spey, mounted on planes recently purchased for TAROM, to develop Romanian technical skills.

> The proposal of repairing in Romania the RR Spey engine (the civil version, mounted on the BAC 1-11) as a preparatory step towards producing an engine of the same category, is considered unacceptable, as it would delay unjustifiably the programme for production of the engine and it would be neither appropriate nor economical, insofar as BAC 1-11 engines are new and not numerous.[44]

In France, Dassault and SNECMA (Société nationale d'étude et de construction de moteurs d'aviation) welcomed the Romanian delegation much more warmly, arranging a special flight to Bordeaux for them and allowing them to visit the facilities there. Discussions with the French companies regarding the manufacture under licence of the modern Mirage F1 and its engine, the ATAR-9K, convinced the Romanian leadership to try and pursue this option. However, the lack of consideration for the SNECMA engine, less powerful and less efficient than the Spey, meant that the question of the engine would remain open, whereas it was decided that the Soviets would continue to be approached along with China, which had started to produce its own version of the MiG-21.[45]

---

41 Michael Hartley, *The Rolls-Royce Spey*, RR Heritage Trust Technical Series No. 10, Derby, 2008; Kenneth Fulton, "Speys International", *Flight International*, 16.11.67, p. 805-811.

42 Bennett – Hamilton (eds.), *Britain and the Soviet Union, 1968-72, op. cit.*, doc. 10, p. 47.

43 Benn, *Office Without Power, op. cit.*, p. 99-100.

44 ANR, CC, Chancellery, 1968_211: notă asupra activității desfașurate de delegația română care s-a deplasat în Anglia și Franța, 18.10.68, f. 94.

45 ANR, CC, Chancellery, 1968_211: protocol No. 63/1968 al ședinței Prezidiului Permanent al CC al PCR din ziua de 16.12.68, ff. 2-4; ANR, CC, Chancellery, 1968_213: cancelaria CC al PRC a tov. Verdeț, Stoica, Patilineț și Ioniță, 20.12.68, f. 7.

At this point, developments in Romanian and Yugoslav aviation began to run parallel. Belgrade was very alarmed and did not hesitate to needle the Soviet Union, stating to its Romanian comrades that the hegemonic policy of Moscow was the forerunner of imperialism and that only concerns about the balance of power with the US had prevented the other socialist countries from going the way of the Baltic republics.[46] On February 2, 1969, in a long *tête-à-tête* with Ceaușescu, Tito returned to the issue of the Soviet threat, emphasising the view that all military equipment for defence in case of invasion was to be produced in the country. Yugoslavia had made good progress in this direction and was able to manufacture two subsonic jets, the Galeb and the Jabstreb, which were internationally praised, but now it was necessary to focus on a supersonic aircraft and close cooperation with Romania would reduce the effort needed to achieve this. Ceaușescu explained that negotiations had taken place for licensing the Mirage, but these were interrupted because of the difficulties encountered. On the other hand, the Soviets seemed even less well-disposed than the French, and, therefore, contacts with the latter were resumed. Tito said that his country had also had preliminary contacts with France for the same purpose, insisting that Belgrade and Bucharest should collaborate towards this common goal:

> President J.B. Tito showed that if Romania and Yugoslavia wanted to obtain the licence of the Mirage, a partnership and cooperation in this regard could be established, and this would be useful and beneficial for both countries. Comrade Nicolae Ceaușescu agreed in principle to the establishment of a closer Romanian-Yugoslav collaboration for the production of armaments, which both countries needed. In this context, he mentioned that the issue of the acquisition and production of the Mirage could also be examined.[47]

Whereas Tito was precise and insistent, Ceaușescu responded rather generically. Moreover, the licence for the Mirage – which, as Dassault and SNECMA made clear from the start, had to depend on the directives of the French government – was never granted, perhaps because of De Gaulle's exit and the blow dealt by the Brezhnev Doctrine to his idea of a Europe stretching from the Atlantic to the Urals. In fact, Romania's priorities towards Georges Pompidou's France were different: cooperation in the recovery of raw materials from third-party countries, help with obtaining favourable treatment with regard to EEC customs tariffs, support for

46 ANR, CC, Foreign Relations, 1968_171: notă No. 3548, 31.12.68, ff. 31-38.

47 ANR, CC, Foreign Relations, 1969_3: convorbirile intime dintre N. Ceaușescu și I.B. Tito, 02.02.69, f. 82. I thank Dr. C. Stanciu for having provided me copy of this document.

accession to GATT and new funding. In the aviation field, attention was focused on the framework agreement of December 3, 1970 for licensing the production in Romania of Alouette III helicopters and, from 1973, of SA-330 Pumas (the latter a source of concern for the French, who feared trouble from the Americans in relation to CoCom). As regards fixed-wing supersonic flight, there is only one Romanian proposal for cooperation in basic research on aerodynamic and lift problems. In general, the French government noted with concern that Romania, after absorbing more than a third of the Western credits destined to Eastern Europe between 1965 and 1969, had a passive trade balance with almost all its partners and that, therefore, and that it was therefore appropriate to limit exposure to financial risk.[48]

Romanian contacts with the Chinese aviation industry were to have unexpected results, while those with the USSR did not come up to Bucharest's expectations. As mentioned earlier, following some rather tricky negotiations, Moscow granted Romania the manufacturing rights for the YAK-52 trainer,[49] an aircraft driven by a 9-cylinder radial engine. However, cooperation on the supersonic fighter was another matter, as was demonstrated by the difficulty in acquiring the MiG-23, which prompted the Romanian Deputy Head of Military Intelligence to proclaim that 'one pound of aircraft costs more than one pound of gold.'[50]

So, about a year after the decisions taken by the Executive Committee in February 1968, on the one hand, the effects related to the Brezhnev doctrine made it impractical to proceed towards building a supersonic fighter and, on the other hand, they led to close cooperation with Yugoslavia. In spite of Tito's insistence to the contrary, the aim was to collaborate in producing a light subsonic trainer aircraft under the name of JUROM – the initial letters of the two countries. As affirmed by Simion Tătaru, director of the project from 1986, the Yugoslavs, thanks to the

---

48 HAEC, MAEF-43: Entretien du 16.06.70 entre M. Pompidou et M. Ceauşescu; compte-rendu des entretiens franco-roumaines du 17.06.70; HAEC, MAEF-46: entretien entre M. Schumann et l'ambassadeur de Roumanie, 11.12.70; entretien accordé par M. Pompidou à M. Manescu le 18.01.71; MAEF, série 29, sous-série 29, dos. 269: note sur la 3e Commission mixte gouvernementale franco-roumaine, 13.01.71; Annexe III: actions de coopération scientifique et technique proposées par la partie roumaine à l'examen des organisme compétents des deux pays, janvier 1971; note sur la coopération sur le SA-330 avec la Roumanie, 20.04.71.

49 NARA, ADD, RG 59: Bucharest to SecState, tel. 835, 21.02.75; "Notă referitoare la stadiul relaţiilor bilaterale româno-sovietice", May 1977, in *România şi Tratatul de la Varşovia*, *op. cit.*, p. 525.

50 NARA, ADD, RG 59: Bucharest to SecState, tel. 8709, 30.11.77.

experience gained with the Galeb and the Jabstreb, already in 1969 had a preliminary draft that was more advanced than the one developed by ICPAS, so it was decided the former was to be used. In addition, it was decided that the aircraft would be powered by two Viper 632-41 engines with a thrust of 17.7 kN. This meant that, from the point of view of the engine, the JUROM project would be a further stage in the Yugoslav experience.[51] In the light of these decisions, the United Kingdom would be involved in the project as a supplier of components and know-how.

3.2 The Rolls-Royce Viper Engine and CoCom Issues

Providing aeronautical material for what was undeniably a military project involving a Warsaw Pact country, naturally posed a problem in terms of strategic export controls. Until the end of the 1960s, the United States had been to some extent the 'conscience' of these multilateral mechanisms within CoCom. While Europe and Japan actively supported the growth of the interchange with Communist countries for political as well as economic reasons, a large part of American public opinion and the Congress continued to see this as somehow immoral. With the Nixon administration, this state of affairs changed and relationships between private companies and the Soviet Union began to be considered as a useful tool for promoting Détente. Thus, agreements in the field of science and technology were encouraged and companies such as Boeing, Lockheed and General Electric were able start industrial cooperation. This naturally led to tensions within CoCom. The United States, which in 1971 had still ranked last among Western exporters to the Soviet Union, by 1973 leapt to the top of the list, thanks to an increase in the number of CoCom exceptions which extended normal commercial competition to East-West trade. In this context, the delays caused by the cumbersome American administrative mechanism with regard to requests for exceptions – not to speak of refusals – gave partners/competitors the feeling of being the targets of abuse. At the same time, in 1976, the Deputy Director of the CIA showed deep concern about the profusion of technology transfers.[52]

---

51 Simion Tătaru, *31 octombrie 2004. 30 de ani de la primul zbor al avionului de luptă si sprijin aerian IAR-93*, paper sent to the Author on 28.01.2015.

52 CIA Electronic Reading Room, Intelligence Community Staff report, IC 76-2387, 15.09.76, release CIA76-RDP80B01676R002900280006-3; Michael Mastanduno, *Economic Containment. CoCom and the Politics of East-West Trade*, Ithaca, Cornell University Press, 1992, p. 135-137, 156-161, 174-177.

This is the backcloth against which, in March 1969, the British government approved the export of Viper engines for the Yugoslav plane. It cannot be ruled out that this played a role in convincing the Romanian authorities to propose cooperating with Belgrade on a light fighter the following month. However, it would appear that when Maurer was in the United Kingdom in late November 1969, he did not mention the Viper. Indeed, the point at issue was the idea of producing both the HS-748 and its Dart engine under licence in Romania – a project that Benn strongly supported. It was probably during a visit to the Derby factory on November 25 that Maurer's delegation made it clear to Rolls-Royce that the Yugoslav project was turning into a joint effort with Romania. They also requested that the Dart/HS-748 and the Viper/JUROM should be incorporated into one package so as to make it more difficult for the British to refuse.[53]

While MinTech's reaction was positive, that of the FCO was not. The Viper was not subject to CoCom controls, but – as noted by Sydney Giffard, Head of Eastern European & Soviet Department – the transfer of technological know-how to countries of the Warsaw Pact for an openly military project was something unprecedented (if we exclude the possible use of the Viper on the Czechoslovak L-29 Delfin trainer, which never came off.) On the other hand, the ambassador to Bucharest, Laskey, emphasised repeatedly that a refusal would jeopardise all Anglo-Romanian aeronautical cooperation and would once again throw Romania back into the arms of the Soviets. The Yugoslavs, of course, demanded implementation of the transaction already discussed with Belgrade, especially after the visit of Prime Minister Mitja Ribičić, which offered the opportunity to renew interest in the Viper. Finally, a further point to consider was the growing impatience of the British government and industry with CoCom. The FCO concluded that it was not necessary to consult with Western partners immediately – even though the British permanent representative to NATO had recommended that the US should be at least informed of the JUROM project – but to wait for the development of contacts between Rolls-Royce and the Romanians.[54]

---

[53] UKNA, PREM 13/3460: official visit of the Rumenian Prime Minister and Foreign Minister, n.d.; record of a meeting between the PM and Mauer at 11:30 on 26.11.69; record of a meeting between the PM and Mauer at 12 noon on 26.11.69; UKNA, FCO 46/525: Candler to Stephenson, 30.12.69; UKNA, FCO 28/247: FCO to Bucharest, tel. 66, 20.02.68.

[54] UKNA, PREM 13/3560: record of a meeting between the Foreign Secretary and Anton Vratuša on 24.02.70; record of a meeting between the PM and Mitja Ribičič on 25.02.70; UKNA, FCO 69/103: Giffard to Brimelow, 6.02.70; Giffard to Brimelow, 7.02.70; Gallagher to Brimelow, 9.02.70; Laskey to Giffard, 19.02.70; Laskey to Giffard, 26.02.70. UKNA, FCO 46/525: FCO to UKDEL NATO, tel. 45, 06.02.70;

In mid-February 1970, a delegation of the United Kingdom went to Bucharest, where Hugh Conway, Head of Rolls-Royce Bristol Division, was greeted by Marinescu, the Deputy Prime Minister responsible for the aeronautical programme, and by Ispaş, who continued to be the animating force of the contacts at a technical level. The Romanians focused on two main things: the need to establish mechanisms for counter-trade to finance the costs of cooperation with the British and the strong interest in outlining a research programme with Rolls-Royce that would allow Romania to develop its technological skills. Conway seemed favourably impressed by this approach and he informed the British government that Rolls-Royce was available to forge a relationship of long-term cooperation with Romania on the Dart and the Viper, while maintaining control over spare parts and services for all engines installed on all aircraft sold to third parties. In this way, the British company would be involved in any future export and could block any transactions with countries to which CoCom applied special conditions.[55]

Proposals for cooperation between the Romanian Institute for Fluid Mechanics and Aerospace Constructions (IMFCA, the result of the merger of ICPAS with the Institute of Fluid Mechanics of the Academy of Romania) and Rolls-Royce envisaged a wide range of activities covering, in practice, all aspects of a modern aircraft engine factory. Such latitude aroused concern in Whitehall, where it was thought that the best thing would be to concentrate all these activities on a specific project, such as the development of another version of the Viper with afterburner. Finally, some rather restrictive guidelines were laid down, demanding that all cooperative activities should be submitted for government approval through the Ministry of Defence's resident technical officer. Everything concerning CoCom, although not subject to embargo, had to be examined with particular attention and the activities relating to the development of the Viper would have to keep it within the limit of 22.2 kN thrust, beyond which strategic embargo would be applied.[56]

It should be noted that the ability on the part of the supplier industry to monitor the use of the final product by third parties was an important aspect of strategic and commercial risk assessment and already fifteen years

Burrows to FCO, tel. 72, 09.02.70; Waterfield to Cooper, 12.02.70; Garvey to FCO, tel. 119, 19.02.70.

55 UKNA, FCO 69/103: FCO to Belgrade, tel. No. 56, 09.02.70; Laskey to FCO, tel. 41, 18.02.70; Conway to Mason, 18.02.70.

56 UKNA, BT 359/25: Ispaş to Conway, 29.04.70; minute by Mallinson, 27.07.70; Thexton to Smith, 08.10.70.

earlier, when it was a question of the licence for the Gnat for Yugoslavia, a Memorandum of Understanding had been enacted that would give the United Kingdom veto power over exports, as well as explicitly excluding China, North Vietnam, and all Warsaw Pact countries.[57] Now, the control of spare parts and maintenance services by Rolls-Royce (which the latter assured had already been accepted by the Yugoslavs) was considered more than enough by Whitehall and, therefore, on March 19, 1970, it was decided to start the consultation process with the allies, emphasising in particular the importance of this mechanism in terms of security. The consultations would not begin with the United States, on the grounds that this would take too much time, but would involve all parties simultaneously, with Italy leading the pack, since Fiat had participated in the development of the Viper 632, and was, therefore, entitled to certain rights.[58]

Within about three weeks, all CoCom partners gave their consent to the involvement of Rolls-Royce in the JUROM project, except for the United States, which maintained its position until the middle of September 1970, when Washington's unwieldy administrative machine also gave the go-ahead, substantially without posing any objections.[59] Without waiting for the Americans, as early as June 1970, Rolls-Royce had told the Romanians to be ready to start serious negotiations on the Viper, but it would not be at all easy. This is partly because, in the meantime, the terms of the issue had changed with the final eclipse of the possibility of the HS-748 licence and with the emergence on the Yugoslav side of a preference for outright purchase of the engines, which would leave the prospect of building the entire Viper only to Romanians.[60] Furthermore, the technical problems and the increasing costs of R&D to develop the RB.211 engine for Lockheed Tristar were driving the historic British brand towards financial ruin. In fact, it declared bankruptcy on February

---

57 UKNA, FO 371/130549: draft memorandum of understanding concerning security of aircraft and aero-engines, n.d.

58 UKNA, CAB 134/2802: ESC(O)(70)5, Sales of Viper engines to Yugoslavia and Romania and their manufacture under license in those countries, 11.03.70; ESC(O)(70) 3rd meeting held on 19 March 1970; UKNA, FCO 46/525: FCO to UKDEL NATO, tel. 97, 19.03.70.

59 UKNA, FCO 69/103: Burrows to FCO, tel. 168, 25.03.70; CoCom (70) 361, record of discussion of 14 April 1970, 22.04.70; CoCom (70) 980, note by the Chairman, 18.09.70.

60 UKNA, FCO 46/525: record of a meeting of the President of the Board of Trade with Avram held on 10.03.70; UKNA, FCO 69/103: Walsh to FCO, tel. 169, 29.05.70; Laskey to FCO, tel. 204, 11.06.70.

4, 1971, and was nationalised a few months later by the Conservative government.[61]

It is, therefore, not surprising that the negotiations dragged on throughout 1971 and that the financial aspect played a key role. Rolls-Royce's collapse was compounded by the problems affecting BN, meanwhile Romania was revising its policy regarding the importation of military technology for equipping its armed forces for the period 1971-75 in an attempt to curb the decline of public finances, in particular by slashing the number of MiG-21s and turning to China to obtain supplies on favourable credit terms.[62]

In this situation, from both a practical and an ideological standpoint, and especially in the light of recent experience, it was becoming difficult for the Romanians to accept what for Rolls-Royce was pretty basic, namely to include an escalation clause in the Viper contract, whereby the cost of the engine was calculated by including in the current price possible variations in the cost of labour and materials up to 7%, which corresponded to the inflation index indicated by the British government. As the minister without portfolio Lord Drumalbyn pointed out during his visit to Romania in January 1972, the Romanian leadership could not understand why they were being asked to cover British inflation, which they considered an internal factor concerning only London, when their own workers were being asked to make sacrifices in the name of independence and national development.[63]

These obstacles were overcome thanks to patient negotiations, which took place during the first weeks of 1972 with the unfailing help of Pop, who, it seems, was the main source of information available to the British Diplomatic Service. In addition to finding an agreement on the escalation formula, an important change emerged during the negotiations. It appeared that the Yugoslavs had also resolved to obtain a licence for the Viper and thus, on February 2, 1972, in Belgrade, the heads of agreement between the three parties involved – the Yugoslav and Romanian government agencies and Rolls-Royce – were signed.[64]

---

61 Pugh, *The Magic of the Name*, *op. cit.*, p. 138-148.

62 ANR, CC, Chancellery, 1971_8: notă No. SZ00188, 28.01.71, ff. 216-223; notă privind reducerile volumenor de import aprobate pentru sectorul de apărare pe anii 1971-1975, 02.02.71, ff. 237-238; protocol No. 2 al şedinţei PP din ziua de 08.02.71, ff. 3-8.

63 UKNA, FCO 28/1532: Laskey to FCO, tel. 353, 22.10.71; Watts to Bucharest, tel. 8, 9.12.71; UKNA, FCO 28/1978: note by Lord Drumalbyn, 20.01.72.

64 UKNA, FCO 28/2006: Laskey to FCO, tel. 5, 12.01.72; Frith to Orford, 18.01.72; Bucharest to FCO, tel. 2, 07.02.72.

The situation was, however, further complicated by different concerns about the JUROM project on both the Romanian and the British side. In Romania's case, the recent debacles in the aeronautical sector pushed Bucharest to ask the British government in London for a kind of guarantee to ensure that the industrial enterprises involved, especially Rolls-Royce, would actually be able to keep the commitments they were undertaking. In Britain's case, the JUROM project was going to involve other British businesses: Marconi for altimeters and pathfinders, Martin Baker for ejection seats and Dowty Rotol for the hydraulic system and the undercarriage. In particular, Dowty, after frustrating negotiations with the Yugoslavs, believed they had, however, a real chance of snatching the Romanian order from their French competitor Messier, but they were under pressure from Bucharest to rescue BN, as well as to help the Romanians to design an afterburner for the Viper. Of course, Dowty realised that this last point, in particular, could create many problems. In light of the increasing number of contacts at industrial level, the Ministry of Defence (MoD) in the United Kingdom began to reconsider the green light given in 1970, narrowing the interpretation of CoCom's intended commitments. Rather paradoxically, this applied not to the engine, but to ancillary equipment.[65]

Although a guarantee from the British government implying Rolls-Royce's unreliability as an industrial partner was out of the question, the FCO believed that – considering the situation in which the British aviation industry found itself – it was necessary to do everything possible to make sure that the ongoing negotiations came to fruition. Hence the reservations towards the attitude adopted by the MoD, along with the belief that the Romanian interest in the afterburner was genuine and not a ploy aimed at fostering espionage. However, some concern remained about the fact that the involvement of a Warsaw Pact country was likely to reduce the export opportunities to Yugoslavia, a country with which relations had been established way earlier and were not subject to CoCom embargoes. Certainly, differentiation between the two countries looked like an option, but it was not easy to put this into practice, mainly because it was not possible to determine the exact terms of the cooperation that united them in the JUROM project. Furthermore, Bullard, Head of FCO's East European and Soviet Department, was highly sceptical as to whether Yugoslavia and Romania actually possessed enough resources,

65 UKNA FCO 28/2006: minute by Russell, 10.2.72; Russell to Bullard, 11.4.72; Russell to Downey, 18.04.72; UKNA, FCO 69/294: Abbots to Wilson, 04.02.72; minute by Sparrow, 17.02.72.

experience, talent and determination to make it happen. According to Bullard, in particular, Ceauşescu's Romania was losing political relevance in a rapidly changing international situation and, consequently, an eventual involvement in the JUROM project required careful evaluation.[66]

The project was organised along parallel lines, one in Yugoslavia and one in Romania, and it was coordinated through regular meetings at various levels. The Romanians were responsible for some parts of the fuselage, the empennage and the tanks; the Yugoslavs were in charge of other parts of the fuselage and the wings. Each country was to produce a prototype, through a constant exchange of information, autonomously providing the equipment – if only because Romania had some constraints of compatibility with ground equipment, which was, of course, standard for Warsaw Pact countries. So, however close cooperation between the Romanian and Yugoslav air force was, to the point of almost becoming a model for expansion to other industrial sectors, the United Kingdom could differentiate between the two countries.[67]

In London, however, in response to the position taken on by the MoD and in order to address complaints from Martin Baker, which was afraid of being refused an export licence because of the presence of Romania precisely when the long negotiations with Yugoslavia had gone through, it was decided to reflect globally on the issue, starting from the specific case of the ejection seat, excluding at first differentiation between the two countries. Instead, Whitehall's reflections should have concerned the definition of general guidelines that could be applied to individual cases of equipment that would be used in a military project of a Warsaw Pact country. This would have altered the principle according to which exceptions to the CoCom embargo should have a civil end-use and user. Only later would the British have a real chance to differentiate between Yugoslavia and Romania, as they had done with the general guidelines.[68]

In late April 1972, the MoD Security Export Controls Working Party acknowledged that from the information available at that time it was not possible to determine whether JUROM was really a joint project and,

66 UKNA, FCO 28/2006: minute by Burns, 10.3.72; Burns to Bullard, 18.4.72; Bullard to Burns, 19.4.72; Thexton to Downey, 25.04.72; Russel to Thexton, 09.05.72; Gil Bennett and Keith Hamilton (eds.), *Détente in Europe, 1972-76*, doc. 57, in DBPO, Series III, Vol. 3, London, Frank Cass, 2001, p. 277-284.

67 ANR, CC, Foreign Relations, 1972_41: stenograma convorbirilor oficiale purtate cu prilejul întîlnirii dintre tov. N. Ceauşescu şi I.B. Tito în ziua de 17.05.72, ff. 2-6; Tătaru, 30 de ani, *op. cit.*, p. 2-4.

68 UKNA, FCO 28/2006: Whitby to Facer, 18.04.72; UKNA, FCO 69/294: Aldridge to Jones, 13.03.72; minute by Facer, 11.04.72; minute by Sparrow, 12.04.72.

consequently, if it was possible to differentiate between Yugoslavia and Romania. Compared to 1970, the British involvement had increased and now specific military equipment was required by Romania, such as the Ferranti ISIS S302 gyroscopic pointing system, for which export had previously been banned. Initiating a process intended to last less than two months, the Working Party concluded that it would be possible to ask for exceptions to the CoCom embargo for military equipment, provided that this did not influence the strategic balance and that it would be possible to justify the individual cases with the political advantage of Romania's autonomy from Moscow. Given that the information received from Belgrade indicated that equipping the Viper with an afterburner was not, at the time, a real possibility, the assessment of JUROM's strategic impact remained the one carried out in 1970. Therefore, on June 7, 1972, the Defence and Overseas Policy Committee decided to continue to support the project.[69]

Convinced that JUROM would never see the light of day, Bullard insisted that it was possible to try to differentiate between the two countries, perhaps unaware of the fact that Romania had somehow joined a project originally conceived by the Yugoslavs. Basically, it was necessary to ascertain Belgrade's reaction to the idea of providing ancillary equipment of the plane on a bilateral basis, something that would definitely favour Yugoslavia, a country not subject to the CoCom embargo. However, when issuing instructions for the request for Martin Baker ejection seats to be presented to CoCom, the Security Export Controls Working Party of July 26, 1972, insisted that it was not possible to differentiate and that, consequently, all inquiries relating to JUROM should be considered to involve Romania as well. In this way, ejection seats could not be supplied to Yugoslavia until CoCom gave the go-ahead for them to be exported to Romania; this was obtained in the second half of September. The same directive was also observed two months later, in December 1972, when it came down to on-board instruments provided by Smith Industries.[70]

It is likely that this attention towards CoCom derived at least in part from the difficulties that departments in Whitehall encountered in determining what represented a problem in terms of exports, as shown, for

69 UKNA, CAB 148/128: DOPO (SE) (72) 6, Yugoslav-Romanian Aircraft project – JUROM, 05.06.72; DOPO (SE) (72) 4th meeting on 07.06.72; UKNA, FCO 69/294: SX/M (72) 3, minutes of a meeting held on 26.04.72.

70 UKNA, FCO 28/2007: Bullard to Brimelow, 25.07.72; Bullard to Thexton, 26.07.72; UKNA, FCO 28/2008: note by Burns, 02.08.72; SX/M (72) 5, minutes of meeting held on 26.07.72; CoCom (72) 2057, record of discussion on 19.09.72. Wright to Abbott, 10.10.72; UKNA, FCO 28/2009: minute by Jones, 01.12.72.

example, by the doubts only eventually advanced by the DTI about the exportability to Romania of machine tools necessary for the production of the Viper, for which the sale of the licence had been negotiated for over two years.[71] More generally, the discussions in Whitehall were part of a broader picture of multilateral negotiations aimed at revising the CoCom embargo list to make it less restrictive. In the course of 1972, in fact, it was possible to overcome America's initial reluctance and address the problem of the so-called 'China Differential'. This referred to those criteria applied to China, Albania and Asian Communist countries, following the Cultural Revolution and the vicissitudes of the war in Vietnam, and which were more stringent than those for the USSR and the countries of the Soviet bloc. However, it was impossible to reach an understanding in areas such as computers and telecommunications – partly because the Europeans feared that a greater openness would play in favour of American corporations.[72]

However, aeronautical materials were not actually top of the agenda and there was a general climate of mistrust with regard to both the utility of the CoCom mechanism and the actions of partners. It was these factors that were crucial in convincing Whitehall to continue supporting the JUROM project.[73] The recommendation of the Defence and Overseas Policy Committee to maintain privileged contacts with the Americans and the French on the issue, rather than sticking to the rules, was mainly due to the fear that – in view of the difficulties that Anglo-Romanian aeronautical relations were experiencing – these partners could exploit CoCom in order to undermine from the United Kingdom's market position.

Meanwhile, in Whitehall, discussions ran parallel, though with a slight difference in timing, to the Rolls-Royce direct negotiations with the Yugoslavs and the Romanians. On May 17, the Memorandum for the Viper licence was signed. Under the agreement, the British company undertook to transfer the necessary know-how so that, over a planned cooperation of 12 years and with a turnover of 63 million pounds, the two Communist countries came to produce the engine independently starting from 1980. The Vipers for the two prototypes of the aircraft would be provided directly by Derby; the following engines would first be assembled using British materials and then, gradually, produced independently. Rolls-Royce also recognised the desire of Romania and Yugoslavia to cooperate with each other and undertook not to exercise restraint in this regard, even though both parties were directly and independently licensed. In this way,

71 UKNA, FCO 28/2008: Burns to Frith, 11.08.72.

72 UKNA, FCO 28/2010: The CoCom List Review, diplomatic report 478/72, 13.11.72.

73 UKNA, FCO 28/2008: Abbotts to Wright, 10.10.72.

should cooperation be interrupted, each party would still be able to obtain supplies directly from Rolls-Royce. In the end, both countries obtained the non-exclusive right to sell the engine or aircraft to third parties with the exception of countries subject to UN embargo and specific cases that would be negotiated between governments.[74]

This last point was the subject of serious litigation. Indeed, the control of exports to third countries was a delicate point in the negotiations for the licence of the Gnat, later aborted. In the case of the Galeb and the Jabstreb, the British government had requested that Rolls-Royce should be told who the final customer was with adequate notice and that the company should have a direct contractual relationship with the customer with regard to product support and spares. This would enable Whitehall to give its consent to the transaction – which could not be applied to countries placed under the CoCom embargo – and, through Rolls-Royce, some form of control over the use of the aircraft would be maintained. Now, the same mechanism would also have to be applied to JUROM and, on this basis, it had been decided in 1970 that the project should be supported.[75] However, in June 1972, the Yugoslavs put forward a draft of the intergovernmental Memorandum of Understanding to the British in which – with reference to the agreement with Rolls-Royce – exports were excluded only towards countries under UN embargo, without alluding to those of the 'China Differential.'[76]

In the climate of intolerance due to political restrictions on trade, the Yugoslav position – which effectively subverted one of the key points on which the British government's assessment was based – was not seen as a possible breaking point, but rather as a solution that was no worse than any other. Rolls-Royce refused to include clauses in its contract that reflected government policy rather than industrial interest. Furthermore, it was believed that JUROM would not be available before 1980, when its market appeal would be low. Moreover, it would still be possible to exercise control by means of the DTI export licences for those critical components that Rolls-Royce would continue to provide at least until the end of the decade.[77]

---

74 UKNA, FCO 28/2007: Russell to Reed, 23.05.72; UKNA, FCO 28/2007: protocol concerning general terms of cooperation, 17.05.72; minute by McKearney, 28.06.72.

75 UKNA, FCO 28/2346: Frith to Lolič, 05.03.70; Thexton to Tallboys, 30.01.70.

76 UKNA, FCO 69/295: Šušnjar to Suffield, [12.06.72].

77 UKNA, FCO 28/2007: Lloyd to Frith, 15.06.72; minute by Sparrow, 04.07.72; UKNA, FCO 69/295: minute by Sparrow, 25.07.72.

The amazing thing is perhaps that, in the rush to sign the Memorandum of Understanding by September 15 (the day the Rolls-Royce licence came into force), combined with the approximate nature of the instructions for the British representative, it was the MoD itself that accepted an extremely watered-down version of the article regulating London's right to intervene in the event of possible future exports. This would no longer be subject to previous consent, but would merely allow the possibility of consultations when one of the parties found it necessary. This laxity reflected British exasperation with a document that, at least in part, acknowledged Romania's earlier request for a governmental guarantee of cooperation with Rolls-Royce and it was immediately picked up by the FCO. Bullard said that it was necessary to send a letter to Belgrade affirming the need to be informed, should a transaction with one of the four 'China Differential' countries arise. In particular, in the case of a similar document with Romania, the formulation would have to be much more stringent and explicit.[78]

This demand was of course bound to create problems. On the one hand, the British found themselves negotiating a Memorandum of Understanding that was as close as possible to that concluded with Yugoslavia, not only by virtue of the equivalence of the two countries in the JUROM project, but also to repel Romania's demands for a written statement from London guaranteeing the counter-trade quota and Rolls-Royce's collaboration – both considered inopportune by the British government. For its part, the United Kingdom insisted on a formulation of the exports clause specifying the four 'China Differential' countries and, more generally, recognising the United Kingdom's substantial veto power.[79]

Thus, the triumphant visit of the President of Rolls-Royce, Kenneth Keith, in what the FCO Under-secretary of State, Thomas Brimelow, described as the toughest police system in Eastern Europe, was partially overshadowed by this new obstacle. Keith and Hooker were received for over an hour and a half by Ceauşescu himself, who tickled British pride by saying that Romania did not intend to compete with Rolls-Royce,

[78] UKNA, FCO 28/2008: Burns to Brimelow, 07.08.72; minute by Burns, 24.08.72; Thexton to Burns, 29.09.72; Bullard to Thexton, 23.10.72; UKNA, FCO 28/2009: Steward to Bullard, 16.11.72; UKNA, FCO 69/69/295: Memorandum of Agreement between the Government of the UK and Yugoslavia, 15.09.72; minute by Burns, 13.10.72.

[79] UKNA, FCO 28/2009: Ashe to FCO, tel. 429, 01.12.72; FCO to Bucharest, tel. 538, 14.12.72; negotiations for a Memorandum with Romania on the JUROM Project – Steering brief, n.d.

but wanted to be like a novice in the production of jet engines.[80] It is easy to imagine the pressure that both Romania and Rolls-Royce exerted on the British government to remove this obstacle, but negotiations between Brimelow and Romanian Ambassador Pungan – once described as a 'gangster type' by the First Secretary in Bucharest,[81] but soon to become a very close collaborator of Ceauşescu in aviation matters – did not produce results, as the FCO was concerned to protect itself against possible complaints from CoCom partners.[82]

The apparent contradiction of the British negotiating position, the pressures behind the scenes from Rolls-Royce and those exercised in Bucharest by those who, like Ispaş, prioritised the successful outcome of the negotiations, finally prevailed. Unsurprisingly, the solution followed the example of the Memorandum with Yugoslavia, with a somewhat laconic clause on exports, allowing the possibility of consultations and excluding only countries placed under UN embargo, adding, however, an exchange of letters which, while not explicitly naming the countries of the 'China Differential,' referred to special cases covered by the political commitments made by the British government elsewhere. After overcoming the resistance at the last minute of the Romanian Foreign Ministry, who considered this formula too restrictive for Bucharest's foreign policy, the Memorandum was signed on March 8, 1973 by the Deputy Foreign Minister Vasile Gliga.[83]

3.3 The Chinese Spin-off

In his memoir, Hooker remembers how, thanks to the diplomacy of Pop, it had been possible to gradually get in touch with the leaders of the Romanian nomenklatura, and how this had had a very significant role in the successful conclusion of negotiations on the Viper. Among the contacts secured by Pop, there was Emil Bodnaraş, Vice President of the Council of State and a top communist dignitary who had been one of the

80 UKNA, FCO 28/2009: record of a meeting with the chairman of RR on 28.12.72; Hooker, *Not so Much of an Engineer*, *op. cit.*, p. 202-203.

81 UKNA, FCO 28/255: Hanbury Tenison to Rhodes, 27.02.67.

82 UKNA, FCO 28/2009: Sir Brimelow's meeting with Mr. Pungan on 15.12.72; UKNA, FCO 28/2010: record of a meeting with Pungan on 21.12.72; UKNA, FCO 28/2346: minute by Green, 22.01.73.

83 UKNA, FCO 28/2346: Ashe to FCO, tel. 68, 02.02.73; minute by Green, 05.02.73; Ashe to FCO, tel. 90, 08.02.73; Ashe to Home, 13.03.73; Ashe to FCO, tel. 156, 08.03.73.

protagonists of at least two relevant moments in the history of postwar Romania: the withdrawal of Soviet troops from the country in 1958 and the rise of Ceauşescu after the death of Gheorghiu-Dej. Hooker speaks of a very friendly relationship with the almost septuagenarian Bodnăraş and how he had to visit his house.[84]

On the evening of February 18, 1972, while the Vice President described the impressions gained during his many trips to China, the Technical Director of Rolls-Royce interjected, saying that he himself would like to go there. In 1936, in fact, the young Hooker had answered an ad for the new Professorship in Aeronautics at the Pei Yang Technical College in Tien Tsien, but the Sino-Japanese War had forced him to scrap the project.[85] It would then appear that both China's argument and what Hooker said were the result of chance, small talk during a pleasant evening. Actually, months before, the Technical Director of Rolls-Royce had already expressed the same desire to a Chinese delegation visiting Bristol and he was obviously seeking Romanian help to achieve it, believing that China was in a situation similar to that of Romania regarding the production of jet engines and, therefore, that the latter could, in turn, be interested in the Viper.[86] Bodnăraş, who was an important *trait d'union* between the Bucharest and Beijing regimes, remarked that it would be a pleasure to point out to the Chinese comrades, at the highest level, the usefulness of inviting Hooker and that they would not fail to do so.[87] Three months later, a Rolls-Royce team left for the PRC.

Although it is difficult to assess the role played behind the scenes by the influence of Bodnăraş in facilitating aeronautical Anglo-Romanian relations, starting from the sale of the BAC 1-11s in 1968, it is true that his actions were considered, *ex post*, as of great importance also by Whitehall officials and Pop, after the Vice President suffered a heart attack, hastened to cultivate his friendship with the Deputy Prime Minister of the Machine Building Industry, Gheorghe Oprea,[88] so that the British aircraft industry did not find itself without adequate support at the right level. Bodnăraş's personal bond with the United Kingdom was represented by his son Gheorghe, who in those months was completing

84 Hooker, *Not much of an Engineer*, *op. cit.*, p. 201, 204.

85 *Ibid.*, p. 22.

86 UKNA, FCO 14/1001: minute by Gillmore, 28.04.72.

87 UKNA, FCO 28/2006: Ashe to FCO, tel. 64, 22.02.72.

88 Oprea was invited on a private visit to the UK by Keith in July 1975. See ANR, CC, Economic Section, 1975_35: Keith to Ceauşescu, 23.07.75, ff. 92-93; notă H.3910, 04.09.75.

his studies in Aeronautical Engineering at Imperial College. This was a connection on which BAC, Rolls-Royce and Dowty – all represented by Pop – had an interest in acting upon, especially after all the recent failures, funding a scholarship that would allow Gheorghe to continue with a PhD and inviting his father to the United Kingdom for a private visit in June 1972.[89]

Precisely at a time when the British government had concretely begun to suspect that the ambitions of the Bucharest regime had taken a megalomaniac bent, negotiations on the JUROM opened unexpected business opportunities that the benevolence of Bodnăraş, brought to London on a Rolls-Royce private jet, would help to bring about. On the other hand, even if these unexpected opportunities increased the immediate importance of Romania for the British, they also allow to put into perspective the relative importance of the Anglo-Romanian relationship when compared with those held with other interlocutors. In fact, the visit to China of the Rolls-Royce team was a great success: the Beijing government showed interest in a series of projects, in particular for a manufacturing licence for the Spey engine, giving Rolls-Royce a chance to resume its penetration into the Chinese market once again, which had begun at the turn of the 1950s and 1960s, only to be interrupted by the political effects of the Vietnam War and the cultural revolution on the fate of the China Differential. Hooker was convinced that the European aeronautical industry had no chance to compete head-to-head with the American one: the Spey, in its military version, thus became a toll to pay to ensure Rolls-Royce a major outlet in China in the years to come. In this framework, Romania had – according to the Technical Director – not only a value as such, although the country represented a smaller market, but also for the close relationship it had with the Chinese, who used it to test the waters and clarify their position.[90]

Indeed, the action undertaken by Romania was not insignificant. Not only did the Chinese invitation arrive on time, after the conversation between Hooker and Bodnăraş, but the Romanian diplomatic staff was also actively involved, in some sort of remote mediation, in the May 1972 visit and in subsequent ones, on which occasions the Rolls-Royce representatives never failed to show lasting gratitude for the role played by Bucharest in setting up the best possible contacts with the authorities

89 UKNA, PREM 15/1185: Alexander to Bridges, 01.06.72; Grattan to Bridges, 23.06.72; UKNA, FCO 21/1395: minute by Cartledge, 02.05.75.

90 UKNA, FCO 28/2006: Peking to FCO, tel. 471, 19.05.72; UKNA, FCO 28/2347: Russell to Bullard, 25.06.73; Engel, *Cold War at 30,000 Feet*, *op. cit.*, p. 187-251.

in Beijing.[91] Hooker, speaking in August 1975 with the British military attaché in Beijing, reiterated explicitly this last aspect, while also recalling how, during the long negotiations, in several delicate circumstances, a word whispered into the ear of the Romanians had helpfully been transmitted to the Chinese with good results.[92] Of course, Bodnăraş had not been driven to act only by the prospects of his son studying in the United Kingdom, but rather by the ambition – surely shared by the leadership of the regime – that this could be functional to the establishment of a trilateral cooperation in the Anglo-Romanian-Chinese aviation industry, from which Bucharest stood to gain large benefits, both political and practical.[93]

This would have been possible if Romania had been a necessary intermediary – and not just occasionally useful – to deal with the Chinese. Only in that case, in fact, could the constraints arising from being a member of the Warsaw Pact be overcome, either through the British willingness to force the limits set by CoCom on the grounds of the political and commercial importance of this mediation, or else, through getting from the Chinese what they had got from Western countries, but that the Romanians were still denied. Naturally, neither situation was likely to occur. In fact, as soon as a direct channel with the Chinese was established thanks to the visit of the delegation led by Hooker in the spring of 1972, both the Ambassador in Bucharest, Ashe, and the Whitehall officials were quick to point out that Romania, at that point, had no useful role anymore. On the eve of the trip to Bucharest of the President of Rolls-Royce in November 1972, Brimelow emphasised to him that since China had begun to deal directly with Western countries and the China Differential had been overcome by CoCom, Romania would no longer be helpful in the future, regardless of whether it had proved so in the past.[94]

Furthermore, even if Beijing had contributed to the development of the Romanian defence industry, the negotiations on the transfer of technological know-how seems to have been rather laborious and – also in light of the subsequent developments – it seems more than doubtful that Romania ever received from China the technical documentation for the construction of a supersonic engine, which had been requested in 1976,

[91] UKNA, FCO 28/2006: Russell to Bullard, 26.05.72; Romulus Ioan Budura (ed.), *Relațiile Româno-Chineze 1880-1974: Documente*, doc. 431, Bucureşti, Ministerul Afacerilor Externe – Arhivele Naționale, 2005, p. 1155-56.

[92] UKNA, FCO 21/1396: Tustin to Stoneham, 27.08.75.

[93] UKNA, FCO 28/2008: Ashe to FCO, tel. 287, 18.08.72.

[94] UKNA, FCO 28/2009: record of a meeting with the Chairman of RR on 28.12.72.

after Rolls-Royce had granted the Chinese the licence to manufacture the Spey 202.[95] So, the very different openness of the United Kingdom towards China – whose epitome is the ready support of the FCO to the idea of licensing the Spey despite a very likely strong resistance at CoCom[96] – only served to highlight even more the constraints imposed on Romania's ambitions because of its being a member of the Warsaw Pact and its offering a much less desirable market potential.

Here it is not possible to analyse in detail the vicissitudes of the complicated Anglo-Chinese negotiations, but it is necessary to consider at least some of them, as they are crucial to understand, in a comparative perspective, the political and economic factors that affected the technology exports to Communist countries and how those factors were not playing in favour of Bucharest.

When the Chinese made it clear that they were interested in the military version of the Spey – an engine clearly subject to embargo for power and employment – the president of Rolls-Royce pointed out to Prime Minister Heath that that transaction was strategic for the company: the Spey 202 could be the key to solve, in the short term, the liquidity problems of Rolls-Royce and to ensure, in the long run, the presence of the British aerospace industry in China. Whitehall officials immediately responded to the point:

> From the industrial standpoint Spey business [...] would be extremely profitable, and it would provide at an early date the liquid funds to expand [Rolls-Royce] research programme, develop new engines and reduce their present heavy dependence on the RB.211. In brief, the deal would give them an opportunity, unlikely to recur, of reducing their dependence on public funds and restoring their commercial position.[97]

The obvious positive results in terms of economy and domestic (even before international) policy deeply influenced the British government and made it much more determined to defend its interests compared to the JUROM case. This does not mean, however, that the bond represented by CoCom was ever ignored. The United Kingdom believed that the

---

95 ANR, CC, Administrative-Political Section, 1976_2: Ioniță la Ceaușescu, 21.04.76, ff. 67-70; livrete continînd cererile de tehnica militara, utilaje pentru industria de apărare și documentație de fabricație, din Republica populară chineză, pe anul 1977, n.d., p. 73-81.

96 UKNA, FCO 28/2009: SX/M (72) 6, minutes of meeting held on 10.11.72.

97 UKNA, CAB 148/138: DOPO (SE) (73) 5, Sale of military equipment to China, 16.04.73.

multilateral export controls had generally operated well up to that point and that the strategic reasons in favour of keeping the embargo were still solid, as the latter contributed – through the maintenance of Western technological superiority – to balancing the quantitative preponderance of the Communist world. The problem, therefore, was not to abolish CoCom, but to adjust their operation to the reality of Détente: the determining criterion should no longer be the ultimate use of what was exported, but its impact on the security of the 'free world', and the burden of proof had to be shifted from the proponents to those who objected to a transaction. The Spey intended for China thus became also an opportunity to broach to the Americans the broader problem of the CoCom reform, taking advantage of the different attitude about it shown by the Nixon administration, especially after the end of the Vietnam War.[98]

Of course, curbing the temptation to act unilaterally, there was perhaps the fear of American retaliation, which could have put at risk the cooperation between the two countries, primarily on the nuclear deterrent. So, when feedback from Washington officials proved negative, the British government decided to turn directly to Henry Kissinger. The National Security Advisor said that President Richard Nixon and himself looked favourably at the transaction, although the reaction of the bureaucracy would inevitably be negative. Upon this assurance, Rolls-Royce was told to continue the negotiations, but when the British submitted the case to CoCom, the most negative reaction came precisely from the USA. Actually, the instructions to the American delegation to CoCom recommended, as promised by Kissinger, just a mild opposition to the British request; what Washington flatly refused was the reform of the functioning of CoCom, as outlined in London. The British then decided to separate the issue of the Spey from that of the CoCom reform, until – with the approach of a new list review in autumn 1974 – they gave up entirely, considering it more convenient to continue to exert pressure to reduce and manage in a more flexible way the embargo within the established rules, rather than trying to modify them.[99]

It should be noted that repeated informal reassurances from Kissinger were sufficient to ensure that the negotiations between Rolls-Royce and the Chinese went on even without the backing of CoCom, even though

98 UKNA, PREM 15/2223: Reappraisal of CoCom, n.d. [Spring 1973]; UKNA, CAB 148/138: DOPO (SE) (73) 3rd meeting held on 11.04.73.

99 NARA, ADD, RG 59: SecState to USMission OECD, tel. 158059, 10.08.73; UKNA, CAB 148/131: DOP (73) 51, Spey Aero Engines for China, 26.06.73; UKNA, CAB 148/152: OPDO (SE) (74) 2, The Future of CoCom, 13.06.74.

this lack of support was not considered lightly. The British insisted that Washington withdrew its formal reservation on the transaction with China, but Kissinger and Helmut Sonnenfeldt were convinced that, in doing so, a precedent that could have led to the dissolution of CoCom could ensue – not to mention the risk that open support to the sale of the Spey to China could inflict damage upon the Détente between Washington and Moscow. Therefore, the two refused the offer.[100]

Meanwhile, the continuation of the negotiations with China and the return of the Labour Party to government, prompted the Rolls-Royce President to address the issue with Foreign Secretary Callaghan. The conversation between the two is interesting because it highlights, once again, the usefulness of Romania as an intermediary for the British company:

> Sir Kenneth Keith explained that it was vital for the Company's cash flow plans and general strategy over the next few years that the deal to sell Speys to China should go through. [...] Once the new Government had approved the export of the Spey engines, Sir Kenneth Keith would, through his contacts with the President of Rumania, be able to convey a message to Chou En-Lai to the effect that the deal was clear from the United Kingdom point of view and now awaited a decision on the part of the Chinese.[101]

Within a few days Prime Minister Wilson gave the green light, so that Callaghan could tell Keith that the British government was anxious to see the negotiations end successfully and that Rolls-Royce could let the Chinese know that they enjoyed their full support.[102]

One of the thorniest problems of the last phase of the negotiations was to convince the Chinese to agree to a re-assignment clause, according to which the latter could not sell, hire or lease engines produced under licence, or transfer its technological know-how to third parties. Such a clause would have, first of all, political significance, in the sense that it would effectively manage the impact of the transaction on CoCom – for example, by assuring Tokyo or Seoul that the PRC would never provide North Korea with military aircraft powered by the Spey – while at the

---

[100] UKNA, PREM 15/2223: record of a conversation between the Secretary of the Cabinet and the American Secretary of State on 30.01.74; UKNA, CAB 148/138: DOPO (SE) (73) 9, Sale of Military Spey to China, 07.09.73; UKNA, FCO 69/529: Washington to FCO, tel. 613, 18.02.74.

[101] UKNA, FCO 69/530: note of a lunchtime discussion between the Secretary and Sir Keith on 13.03.74.

[102] UKNA, FCO 69/530: Bridges to Acland, 22.03.74; record of a meeting between the F&C Secretary and Sir Keith on 25.03.74.

same time substantiating the informal reassurances given by Kissinger. Considering that Rolls-Royce did not seem to have competitors in China, Whitehall had, however, decided that it was up to the British company to 'cover up' for the government, insisting with the Chinese that Rolls-Royce could not give – for purely commercial reasons – a licence that would make them competitive in third countries. The possibility to proceed as in the case of JUROM – that is, through an intergovernmental Memorandum of Understanding that provided for consultations in case of possible exports – was shelved for the moment, assigning to Rolls-Royce the task of negotiating with the Chinese what, at the time, the FCO considered a fundamental political point.[103]

The Rolls-Royce Board agreed to insist on the re-assignment clause, but they made it clear that they did not consider it really necessary, as far as the interests of the company were concerned, if not because, ultimately, its concrete application would still heavily depend on China's good will. So, when the Chinese made it clear, in September 1975, that they were unable to accept a contractual clause which bound their freedom of action, considering that it would have represented an unacceptable precedent from the political point of view, Keith began to suggest that he was not prepared to put at risk the closure of the negotiations for something that was mainly a problem of the government. Whitehall officials, in fact, were convinced that it was necessary to state that the consequences of the Anglo-Chinese transaction would be limited – but this was considered impossible without the re-assignment clause, since it was still unclear on what type of aircraft the engine was to be mounted –, and that the Romanians had discreetly tested the waters several times, showing their interest in the Spey 202. Fortunately, China had decided to enclose the clause, if not in a proper Memorandum of Understanding, at least in a side letter, which the contract should not refer to. Obviously, while being willing to accept the side letter, possibly also with a formulation of compromise milder than the original, the British government decided not to communicate it to Rolls-Royce, so that the company would continue to insist, at least until the moment when it became evident that the negotiations were coming to a standstill.[104]

---

103 UKNA, FCO 69/532: Peking to FCO, tel. 1000, 17.12.74; UKNA, FCO 21/1393: summary of meeting held on 03.01.75 to discuss Speys for China.

104 UKNA, CAB 148/152: OPDO (SE) (75) 8, Speys for China: re-assignment clause, 09.09.75; OPDO (SE) (75) 3rd meeting held on 12.09.75; UKNA, FCO 21/1396: Peking to FCO, tel. 688, 09.09.75; UKNA, FCO 21/1397: minute by March, 18.09.75.

The fears and hesitations in Whitehall should once again be read against the backdrop of the frustrating and concomitant multilateral negotiations of the CoCom review list. After a promising start, the Ford Administration had had second thoughts about the degree of liberalisation of the embargo, showing at the negotiating table a mixture of intransigence and lack of preparation, which created much irritation. Even if it is difficult to assess the direct impact of the Spey on this negotiation, the British were of the opinion that the determination of the United States officials not to make concessions in the aerospace sector derived from the original suspicion that the requests to redefine the operation of CoCom had been instrumental in the cession of the manufacturing licence to China. In this context, and in the face of what was considered America's objective difficulty in formulating a position shared by all the members of the Administration, the British once again asked Kissinger for confirmation about the repeated reassurances received in the previous years, in particular expressing the fear that the United States would raise new issues in CoCom. The Secretary of State robustly answered that he had no objection to the conclusion of the transaction and that, as far as he was concerned, it did not matter what the other departments in the Administration might recommend.[105]

With this additional reassurance and the negotiations in Beijing almost nearing their end, on November 12, 1975, Frank Beswick, Minister of State at the Department of Trade and Industry and the future first President of British Aerospace, recommended dropping the re-assignment clause, should it ultimately affect the closing of the agreement, perhaps exploiting this concession to impose on the Chinese a higher fee for the licence. This proposal provoked immediate positive results and, within three days, Wilson gave the go-ahead. Given this general consensus, which most likely represents the moment in which the British most significantly departed from the multilateral constraints CoCom, the compromise solution offered by the Chinese was immediately accepted on November 14. According to this, restrictions on the use of the licence would last ten years – as opposed to being unlimited – and the commitment not to transfer the engines outside the territory of the PRC would be recorded in a side letter to be penned on the very same day of the signing of the contracts.[106]

105 UKNA, CAB 148/152: OPDO (SE) (76) 1, CoCom: the 1974-75 list review, 09.02.76; UKNA, FCO 21/1397: Washington to FCO, tel. 3292, 10.10.75.

106 UKNA, FCO 21/1398: Beswick to Hattersley, 12.11.75; note by Popplewell, 13.11.75; Peking to FCO, tel. 930, 14.11.75; Wright to Akenhead, 15.11.75; FCO to Peking, tel. 782, 14.11.75.

Of course, problems surfaced until the end, to the extent that the celebratory reception was held before the signing to allow delegations to resolve their remaining differences, but on December 13, 1975, the negotiations were successfully completed. Rolls-Royce obtained 41.5 million pounds for the sole manufacturing licence, more than expected a few months before, and the Ambassador to Beijing, Edward Youde, was able to properly underline the importance of the agreement finally reached:

> The final contracts are important and profitable for the Company. They are also of great political significance. The Chinese have in effect, despite their policy of self-reliance, based the development of their aero-engine industry on collaboration with a British firm. [...] For the future, Rolls-Royce have won an inside track for the British aero-space industry in China.[107]

An important achievement, at least initially, also thanks to the good relationship with the Romanian government, to which, however, as we shall see, the constraints imposed by CoCom continued to be applied in an ever more stringent way.

3.4 Afterburning

During the negotiations for the sale of the Viper manufacturing licence, Rolls-Royce proved willing to participate, along with the Romanians, in an R&D plan to develop a variant of the engine afterburner. Basically, the British company was to coordinate the programme and provide material and equipment, thereby contributing to the creation of the necessary research and testing facilities in Romania. The afterburner as such was not subject to the CoCom embargo: as Brimelow explained to Keith before the latter went to Romania in December 1972, the important thing was that the engine did not exceed a thrust of 22.2 kN. Were this limit to be respected, there would be no problems from a technical point of view and Whitehall officials were confident that Rolls-Royce would strive for this to happen.[108]

The Romanians, of course, insisted that the development plan should follow suit – persevering on the creation of a domestic industry of aircraft

---

107 UKNA, FCO 21/1400: Peking to FCO, tel. 1100, 15.12.75.

108 UKNA, FCO 28/2009: minutes of discussions on 15.09.71; record of a meeting with the chairman of RR on 28.12.72; UKNA, FCO 69/295: minute by Haynes, 07.07.72; Cox to Frith, 02.08.72.

engines based on Rolls-Royce technology and licences – but the British tried to gain time, albeit the project appeared harmless. The problem, in fact, was not a technological one: it was obvious that the afterburner for the Viper would be almost irrelevant for the strategic balance between the East and the West. The difficulty, if anything, was political and formal, since Whitehall officials did not want to return to CoCom with a proposal that would alter the framework of the JUROM project, as it had been provided in 1970. With all the attention now fixed on the negotiations with China, there was no intention to reopen the JUROM dossier at CoCom, perhaps risking that doing so would negatively impact on the other project, despite the interest shown by Rolls-Royce to participate in the development of the afterburner.[109]

Facing strong reluctance, mostly motivated by the fear of possible embarrassment with Western partners, Ambassador Ashe advised against being bound to anachronistic and illogical positions, especially considering the efforts made to sell the Spey to China, while also recommending that care should be taken not to be less available than the American and French competitors. Ashe warned that, because of what had happened in recent years in the United Kingdom, the British aviation industry had less to offer now than it had in 1968, when the 1-11s had been sold to TAROM. As such, it was particularly urgent to capitalise on the opportunities that opened for Rolls-Royce. Therefore, the JUROM had to be considered the hub of Anglo-Romanian aviation relations not only because, politically, it contributed to Bucharest's autonomy from Moscow, but also because it would allow the United Kingdom to consolidate its role as Romania's primary partner in that sector.[110]

However, Ashe's analysis failed to grasp one important aspect. At that time, admittedly, Ceauşescu's purpose was not autonomy from Moscow *per se*, but rather the development of a national arms industry capable of rapidly freeing itself from licences – be they Western or Soviet –, thus being able to export products at its own will. The ultimate goal, then, was not a partnership with any specific party, but the creation of a production base that was to consolidate Romania's political position and contribute significantly to its economy. To this end, the leadership was not only willing to accept, but also to actively promote a certain degree

109 UKNA, FCO 28/2010: Russell to Green, 22.12.72; UKNA, FCO 28/2346: Smith to Thexton, 23.02.73; UKNA, FCO 28/2347: SX/M (73) 3, minute of a meeting held on 16.03.73.

110 UKNA, FCO 28/2347: Ashe to Bullard, 24.04.73; UKNA, FCO 28/2349: Ashe to Home, 19.03.73.

of specialisation within the Eastern bloc (as in the case of the aircraft after obtaining the Soviet licence to produce the YAK-52), just when this would favour its exports and would help stabilise the trade balance of military products with the countries of the CMEA.[111]

The problem was that things were not going as quickly as Ceauşescu would have liked, both because this task mercilessly exposed systemic deficiencies of the Romanian planned economy, and because excessive voluntarism led to underestimation of the practical difficulties posed by the assimilation of technological know-how. Of course, the tendency was to address these problems by tightening bureaucratisation and centralisation, for example with the creation of the Technical Council for the Coordination of Military Technology Production in 1974.[112] In the case of a relatively sophisticated product like the JUROM, these difficulties increased – not to mention the fact that Yugoslavia's consent would be required to export the plane in the CMEA countries – and the resulting delays pushed Ceauşescu to lose his temper, as early as November 1973, when he stated that 'if we proceed at this pace, we'd better give up. Generally, we're sleeping on this subject.'[113]

The JUROM prototype that flew in October 1974 weighed, in fact, almost a ton more than expected, making it all the more urgent to produce an afterburner that would increase the thrust of the Viper, while also requiring a substantial redesign of the aircraft before it reached the pre-series and series production stages in order to reduce its weight. There were also other obstacles directly related to manufacturing, for example in the production of the fuselage structures by milling. So, if the production schedule adopted after the completion of the prototype required the Romanian line to proceed faster than the Yugoslav one, and to conclude the production of the aircraft about a year in advance, the accumulation of delays meant that, by the end of 1976, Bucharest was forced to conform to the same rhythms as the Yugoslavs.[114] Even if Ceauşescu had ordered, in May 1977, that the problem of the excess weight be solved, while also

---

111 ANR, CC, Chancellery, 1973_159: raport privind proiectul punctului de vedere al delegaţiei române în Comisia permanentă CAER pentru industria de apărare, No. 03.764, 31.10.73, ff. 129-134; protocol No. 35 al şedinţei PP din ziua de 05.11.73, ff. 2-4.

112 Decretul No. 155 din 19.06.1974 privind organizarea Consiliului Tehnic de Coordonare a Producţiei de Tehnică Militară, available on http://lege5.ro/.

113 ANR, CC, Chancellery, 1973_159: stenogramă şedinţei PP din 05.11.73, f. 13.

114 ANR, CC, Chancellery, 1977_50: notă privind necesitatea actualizării dinamicii de producţie a avionului reactiv subsonic de luptă IAR-93, No. 1576, 20.11.76, ff. 18-19; notă No. PS3553, 08.04.77, f. 17.

intensifying efforts to produce the aircraft (in general, his side notes on the long reports describing the growing problems of the Romanian production base merely came down to demanding that measures should be taken to achieve the targets set),[115] the picture that emerged in the second half of the 1970s was that of an economy plagued by a growing incapability to keep up with the ambitions of the political leadership.

Even focusing on the aeronautical sector only, it is possible to understand how similar problems in the assimilation of the know-how transmitted thanks to the licences obtained was being experienced even in the case of the Soviet YAK-52, which was also far less complex to assemble. More generally, what emerges is a picture dotted with material shortages, insufficient skilled personnel, lack of coordination between the authorities and difficulties on the part of technicians to use the documentation received from abroad:

> In 1977, the Ministry of the Machine Building Industry gave insufficient attention to the [manufacturing] of spare parts assimilation programmes, recording huge setbacks, particularly those intended for the use and repair of armoured vehicles and aircraft.[116] The underperformances recorded are determined mainly by the delay in the processing of documentation and in the launch of new investments in most businesses, as well as by the lack of manpower and organisation at the Company's mechanical construction sites in Sadu and at the Enterprise for the Repair of Aeronautical Material [IRMA] in Bucharest.[117]

It should be noted, however, that the problem was not at all the lack of investment: with the notable exception of the aircraft engine factory in Bucharest, other establishments in the aeronautical sector showed difficulty in absorbing the allocations made in their favour. Of course, this did not mean that the production of the engines went on smoothly, but rather it exposed the disproportion between the productive performance of the Romanian economy and the aeronautical ambitions of the regime. As for the JUROM, at the end of the 1970s, delays in the approval process, along with further problems being detected in the stabilisers, made the

[115] ANR, CC, Administrative-Political Section, 1976_2: indicațiile date de tov. N. Ceaușescu cu prilejul vizitării expoziției de tehnică militară, 16.06.77, ff. 91-94.

[116] ANR, CC, Administrative-Political Section, 1978_10: raport privind îndeplinirea sarcinilor pentru anul 1977 din Programul unitar al dezvoltării industriei de apărare, 20.01.78, f. 51.

[117] ANR, CC, Administrative-Political Section, 1978_10: raport privind realizarea producției și asimilărilor de tehnică militară, investițiilor și exportului special pe trim. I 1978, 18.04.78, f. 126.

production schedule yet again uncertain, with the whole project finalised no sooner than 1981.[118]

If the problems with the JUROM (or IAR-93, according to the Romanian designation) were not enough, in 1975 – with the Anglo-Chinese negotiations now entering their final phase – it had become clear that Romania wanted the Spey 202. The signals in this sense had been repeated ever since 1973 and the interest in the Spey probably explains the calmness – back then judged surprising – with which the Romanians had reacted to the British refusal to collaborate on the production of the Viper afterburner.[119] The unofficial reaction of the British government was one of total closure. As Brian Cartledge, Head of the FCO Eastern European and Soviet Department, explained to the ubiquitous Pop, it was something impossible to realise:

> I said that the supply of engines for the military air-arm of a member of the Warsaw Pact did not seem to me to be a very realistic prospect. Mr. Popp [*sic*] said that, as the Romania Vice President [Bodnaraş] had told him very recently, Romania would like nothing better than to leave the Warsaw Pact, provided that she could acquire the means of self-defence from the West, but that it was still impossible for her to do so.[120]

There is reason to believe that such advances of virulent anti-Sovietism in private conversations did not create the commotion they would have caused a few years before, if anything because now the Romanian challenge to the bloc discipline appeared as taken for granted.[121] Surely, this was not enough for the British to flinch, so Bucharest officials were careful to put the issue of the Spey in hypothetical terms. Ceauşescu himself, during Wilson's visit to Romania in September 1975, cautiously stated that Romania, for the moment, had no direct interest in the Spey, but rather wanted to raise the issue of the limits that the CoCom embargo imposed on the possibilities of bilateral economic cooperation. Tactically, it was a clever move because it hit a nerve, so much so that Wilson promptly

118 ANR, CC, Administrative-Political Section, 1976_2: situația realizării produselor noi de tehnică militară cu faze de asimilare în semestrul I 1979, 12.07.79, ff. 116-126; Tătaru, 30 de ani, *op. cit.*, p. 8, 13-14.

119 UKNA, FCO 28/2347: Ashe to FCO, tel. 259, 14.06.73; visit of Sir Keith to Romania on 18-23.10.73.

120 UKNA, FCO 21/1395: minute by Cartledge, 02.05.75.

121 Bennett – Hamilton (eds.), *Détente in Europe, 1972-76*, doc. 67, *op. cit.*, p. 326-331.

reported the content of the conversation to the American Secretary of Defense, James Schlesinger, in order to stress the need to reform CoCom.[122]

Moreover, juxtaposing two things that were very different from the technical point of view – the afterburner for the Viper, which would respect the rules of the embargo, and the Spey 202, whose export to a Communist country should have been prohibited – the Romanians secured the involuntary help of Rolls-Royce in lobbying the British government. Keith, for example, was very clear about the question of the Viper in an interview with Wilson in October 1975:

> I said I thought that this [opposition] is ridiculous as the only people who could object to us selling re-heat on the Viper engine to Romania and Yugoslavia would be the Russians. I expressed the view that CoCom had now outlived its usefulness and we tend to suffer from the fact that the United Kingdom pays more attention to it than our competitors do, which makes our sale efforts that much more difficult.[123]

Ridiculous as it might sound, it is a fact that Romania's membership of the Warsaw Pact made these pressures unsuccessful, at least momentarily. Despite the manifest signs of impatience towards CoCom, even in the summer of 1976, the official position of the British government remained the same: even if it had been willing to go all the way just to see the licence agreement with China conclude happily, it now went back to forbidding the development of the Viper afterburner on the grounds of a commitment undertaken years before with CoCom not to provide materials that would improve the military capabilities of the JUROM.[124] Similarly, as much as Rolls Royce could be interested, in general, in securing new contracts in a rather delicate period, and as much as the company could, perhaps, nurture a sense of gratitude for the mediation action carried out by its Romanian contacts with the Chinese, it had always been clear in ruling out the opportunity of cooperation on military issues without governmental approval.

> Regarding the problem with the afterburner for the Viper engine, the company's [Rolls Royce] delegates have pointed out that the approval of the British government is required to begin discussions and they have taken note of our request, as the corresponding documentation has to be drawn up. At

122 UKNA, PREM 16/636: note of a meeting between the PM and the President of Romania on 16.08.75 at 5 p.m.; UKNA, FCO 21/1397: note of a meeting between the PM and the US Secretary of Defence on 27.09.75.

123 UKNA, FCO 21/1397: Keith to the Chairman's Executive Office, 09.10.75.

124 UKNA, BT 359/9: minute by Wilberforce, 29.07.76.

> the same time, they have showed that the approval of the British government only concerns Romania, as it is not necessary for Yugoslavia. Similarly, the company's representatives have shown that these approvals are also required for the engine [Spey] 202, but they have suggested that, at their return home, they will take care of the necessary procedures.[125]

In the autumn of 1976, things started to become clearer. On the outskirts of Bucharest, the Romanians had built a huge factory for the production of aircraft engines, the Întreprinderea Turbomecanica, whose size clarified beyond any doubt that the ambition was not limited to the manufacture of the Viper under licence. The main machineries, sophisticated and expensive, had been bought in the West – especially in the US, Japan and Western Germany – on Rolls-Royce's recommendation (though these guidelines had not always been followed: for example, Italian broaching machines were bought instead of those of Cincinnati Milacron) and had been installed. In short, everything was ready to start work on the Viper. According to the 1973 contract, the local assembly of the engine should have started as early as 1975, to make way, the following year, for its manufacture which would have gradually led the Romanians to produce the whole engine starting from 1980. In autumn 1976, however, delays and problems accumulated. Rolls-Royce had prepared the kits to assemble five engines and raw materials for local production of seventy sets of modules, but deliveries were first delayed by haggling over price and then suspended due to the Romanians failing to pay. Rolls-Royce had also sent three prototypes and eight production engines, of which six had been sent back to replace the combustion chamber in the face of the overall redesign of JUROM because of its excess of weight. The delays and the difficulties, finally, had resulted in a certain degree of irritation among the British company executives, who, with the sole exception of the Vice President, Donald Pepper, did not conceal a downsized enthusiasm for the Romanian connection.[126]

Thanks to Pepper's actions, between the end of 1976 and the onset of 1977, these problems were solved. Even with the redefinition of the production programme of the JUROM between Romania and Yugoslavia, which involved the construction of 175 aircraft instead of the 200 initially indicated, the demand for engines remained unchanged (500 total units, of which half to be produced under licence in Romania) and Rolls-Royce – whose technicians now could access Turbomecanica – prepared a new training

125 Archive of the Romanian Ministry of Foreign Affairs (RMFA), 1976_528: Pinzaru to Gheorghiu, tel. 03/016936, 10.11.76, f. 13.

126 UKNA, FCO 28/2919: Petersen to Cartledge, 06.10.76; visit to Turbomecanica on 20.10.76.

plan for the Romanian staff. With the Viper programme ponderously put back on track, talks on the afterburner resumed. Since the plane was heavier, more thrust would be necessary to get the same performances. Provided that the engine did not exceed the 22.2 kN limit and that the military capabilities of the JUROM were not substantially altered by the addition of the afterburner, Whitehall officials at last permitted Rolls-Royce to help the Romanians – who had already carried out extensive design work in this regard – especially with consulting functions.[127]

The development of the afterburner met a series technical hitches and a new confrontation between the Romanians and Rolls-Royce followed on the interpretation of the parameters of the combustion chamber, so that, as late as February 1979, Romania had yet to sign the corresponding contract – which had already been drawn up by the Yugoslavs in September 1978 – thus refusing to give financial support for the expenses incurred by the company for the development of the new Viper Mk. 633-47, whose thrust went from the 17.7 kN of the Model 632-41 to exactly 22.2 kN with the afterburner. The technical problem – connected to the weight of the engine – was solved by adding 30 kg of lead on the forward fuselage of the single-seat IAR-93, which, since 1980, began to be produced along with the MK. 633-471.[128]

Banking on the experience of the IAR-93, in the first half of the 1980s, Romania autonomously developed an advanced training aircraft, called IAR-99. Much like the former in terms of specifics, but slightly faster, it was substantially equivalent to the Aermacchi MB-339 and a production of fifty specimens, all destined to the Romanian Air Force, was implemented at the end of 1986. What is worth noting here is how, over fifteen years after the JUROM project and in a period of substantial autarky for the Romanian economy, still about 35% of the materials necessary for its manufacture had to be imported. Even the main components made in Romania by means of industrial cooperation (that is to say, the Viper

127 UKNA, FV 14/86: minutes of a meeting held in the Foreign Ministry building in Bucharest on 06.11.76; note of a meeting at DOI on 15.11.76; minute by Weaire, 21.01.77; UKNA, FV 14/87: minute by Weaire, 08.02.77; Dorman to Gerhard, 22.02.77; UKNA, FV 14/90: brief for Mr. Kaufman's meeting with Mr. Almăşan on 26.09.77.

128 UKNA, FV 14/158: meeting with gen. Bucur on 26.10.78; Pepper to Popa, 15.02.79; UKNA, FV 14/159: Warrington to Farrow, 20.03.79; Tătaru to the Author, 15.07.2015.

and the undercarriage) continued to require a substantial share of import materials.[129]

The IAR-99, therefore, clearly shows the limits, on the technological and production levels, of the process of assimilation initially stimulated by the Brezhnev Doctrine while, at the same time, revealing the importance attributed to the aviation industry until the end of Communism. Discussions on the price to be attributed to the plane, at times frankly grotesque, is the epitome of the sclerosis of the Ceauşescu regime and of how he and his wife Elena considered Romania as a sort of family business to be handled to their liking.[130] Finally, the IAR-99 would be modernised in the avionics in the early 1990s, in an attempt to make it marketable in Africa and Southeast Asia, and it is currently the focus of a study to convert it into a plane to be used in air police missions.[131]

In the fall of 1976, the Romanian interest in the Spey 202 had become more directly manifest. In fact, the plan to build a supersonic fighter had not been completely shelved and now the experience in the design of the JUROM could be used to create an aircraft that, powered by the Spey 202, could roughly become the equivalent of the F.4 Phantom. In October 1976, during the visit to Romania of the Minister of State for Foreign and Commonwealth Affairs, Lord Gorowny-Roberts, after Ceauşescu declared that he was personally following the issue, the Deputy Minister of the Machine Building Industry, Gheorghe Boldur, and the Head of the General Aviation Industry, Vasile Bucur, insisted very much on it, recalling – in the face of Gorowny-Roberts's reluctance – the 'precedent' of the licence granted to China, but mostly harping on the fact that the British openness to act accordingly would be an important factor in the launch of the negotiations on the ROMBAC project, id est, the licensed production of the BAC 1-11 (which mounted the civilian version of the Spey) in Romania.[132]

This was the beginning of a diplomatic offensive backed by the entire Romanian nomenklatura involved in the aviation industry, focusing on

129 ANR, CC, Chancellery, 1986_93: preţul de producţie şi de livrare pentru avionul de şcoală, antrenament şi atac la sol IAR-99, 02.12.86, ff. 10-13.

130 ANR, CC, Chancellery, 1986_93: stenogramă şedinţei Biroului Permanent al Comitetului Politic Executiv din ziua de 05.12.86.

131 Arie Egozi, "IAR-99 Upgrade Flight-tested in Israel", *Flight International*, 28.02.94, p. 15; Simion Tătaru, Constantin Olivotto and Vasile Stefan, "IAR-XT/Next Generation", *INCAS Bulletin*, Vol. 4, No. 1 (2012), p. 125-132.

132 UKNA, FCO 28/2919: note of a meeting between the Minister of State and the Romania Deputy Minister at the Ministry of Machine Building on 20.10.76.

the 'precedent' of the Chinese contract and intended to link the military project to the licence to produce a civilian airliner. Returning from a trip to China, Oprea, a 'friend' of the British, who had been close to Pop even before the death of Bodnăraş, went so far as to outline the contours of a possible Anglo-Romanian joint company for the production of the Spey 202. Pepper was quick to note that, indeed, had Romania already produced the civilian version of the engine to equip the ROMBAC, it would be logical to manufacture the military one as well. However, it would have been counterproductive to subordinate the former to the latter, undoubtedly because the civil engine represented a useful viaticum to master the manufacturing technique of the military engine.[133] So Pepper capsized the relationship imposed by the Romanians with the ROMBAC and left the door open to extend military cooperation in the field of jet engines, though always pointing out – as had already been done with the Viper afterburner – that this possibility depended on the decision of the British government.

> [Rolls-Royce] wished to have more business. He [Pepper] would therefore discuss the matter with his Board and would recommend to HMG that they look favourably on the proposal. But as the Deputy Prime Minister [Oprea] was aware the military Spey was equipment over the sale of which Rolls Royce did not have control itself. Rolls-Royce could make its representations but the Government had to take account of international obligations and political considerations outside the area of responsibility of Rolls-Royce. The decision would therefore rest with the British Government.[134]

When it was announced to the Romanian government that, after years of denial, it would be possible to proceed with the development of the afterburner for the Viper, Deputy Minister for Foreign Trade Constantin Stanciu, replied that the interest of his country focused primarily on the Spey. Thus, Rolls-Royce sent to Bucharest, between February 13 and February 16, 1977, a delegation led by Hooker and Gordon Page, the company's Marketing Manager for Europe, in order to understand what the plans of the Romanians were and if they wished Rolls-Royce to officially ask permission from the British government. During the visit, the warm welcome they received proved that the death of Bodnăraş had no negative impact on the relations between the British company and the Romanian government. In particular, Rolls-Royce was very interested in

---

[133] UKNA, FV 14/86: Dorman to Gerhard, 10.11.76; Petersen to DOI, tel. 119, 29.12.76.

[134] UKNA, FV 14/86: minutes of a meeting held in the Foreign Ministry building in Bucharest on 06.11.76.

the information received: Romania was planning to develop a supersonic single-seat aircraft of its own design, without the involvement of Yugoslavia, while also intending to implement a coordinated programme for the production of both versions of the Spey engine – the civil engine would be manufactured first, to accumulate experience, as suggested by Pepper. The production of the 202 would begin in 1981 to achieve full autonomy in 1986-87 and churn out 250 engines by 1990. If Hooker and Page judged unrealistic the timing of the programme, they found the terms of the Romanian proposals very intriguing, which would guarantee Rolls Royce a huge profit. These positive reactions echoed in the recommendation of the Ambassador to Bucharest, Reginald Secondé, to endorse the line defined by Pepper regarding the link between the two versions of the Spey in terms of manufacturing, as this would probably persuade Romania to buy the licence to produce the 1-11.[135]

Reactions in the United Kingdom, however, were diverse. The Rolls-Royce Board concluded, on the basis of the experience so far gained, that the technology transfer would occupy too many of its best engineers, diverting them from other more important tasks. In addition – and this was the crucial point – the Board greatly feared an adverse reaction from the United States, considering that relationships with the new Carter administration had become far more difficult than in the past on these issues, and that the American market was by far the most important for the company. These same conclusions are reflected in those of Whitehall officials, who noted that the transfer of the licence to China did not have the same implications of a case involving an ally of the Soviets. It would have been wrong to encourage Romanian expectations and dangerous to raise false hopes on the successful conclusion of the ROMBAC negotiations positively influencing decisions on the military version of the Spey. In fact, with Ceauşescu's visit to the United Kingdom scheduled for the following year, it was necessary to clear up any possible misunderstanding.[136]

Certainly, there were arguments in favour of a softer approach to the problem. The technology incorporated in the Spey 202 could no longer be called avant-garde, and it was possible to formulate a licence that would grant far less freedom of manoeuvre than that given to the Chinese. Furthermore, according to the Assistant Director for Technology Sales

---

135 UKNA, BT 359/21: Secondé to FCO, tel. 49, 21.02.77; Secondé to FCO, tel. 48, 21.02.77; UKNA, FV 14/87: minute by Weaire, 08.02.77; Dorman to Gerhard, 22.02.77.

136 UKNA, BT 359/9: Cartledge to Gerhard, 11.05.77; UKNA, FV 14/88: Gerhard to Dorman, 17.03.77.

at Rolls-Royce, Danis Jackson, who had negotiated the agreement with China, the Romanians would never be able to produce more than half of the content of the engine, while always depending on the British company for the supply of essential components. All this without considering that, even when Romania had completed work on the Spey 202, it would have to overcome the formidable obstacle of developing a performing fuselage and endowing it with suitable avionics. Nevertheless, at the meeting of the Strategic Exports Working Party, held on August 24, 1977, the matter was shelved in the face of resistance from the FCO – determined to make sure that the licence to China should not be used as a precedent – and of the open hostility of the Ministry of Defence (MoD), which strongly emphasised that the ROMBAC project would have to be scrapped altogether, should the Spey 202 be invoked by Bucharest as an essential condition to close the deal.[137]

Facing the British opposition, Ceauşescu had ordered the supersonic fighter to be completed by 1985, with the Spey or with another engine.[138] It is very likely that the Romanian interest in the Soviet MiG-23 had to be read against this backdrop. The cautious survey conducted in April 1978 by Prime Minister Callaghan, during a meeting with Pungan, in London to work out the details of the impending state visit of Ceauşescu, and which focused on Romania's dependence from the USSR for military aircraft, seems to have gone in this direction.[139] Moreover, for the Romanians it would be the usual trial and error process, trying to keep all doors open, while leaning on the Soviets or the West, according to convenience and, perhaps, trying to get more advantages by pitting the two sides against one another. Discrete hints at the Spey 202, in fact, would not subsequently subside.[140]

In February 1980, Pop, an untarnished representative to Romania for the British aviation industry, spoke to the British Embassy Commercial Counsellor to Bucharest about the Romanian renewed interest in the Spey 202. With the production schedule of the civilian version now satisfactorily launched, Rolls-Royce had indeed an interest in verifying once more the possibility of selling the military licence.[141] A few months later, Pepper

---

137 UKNA, BT 359/10: minute by Octon, 28.06.77; record of SXWP meeting on 24.08.77; UKNA, FV 14/88: note of a meeting at DOI on 21.06.77.

138 ANR, CC, Administrative-Political Section, 1976_2: indicaţiile date de tov. N. Ceauşescu cu prilejul vizitării expoziţiei de tehnică militară, 16.06.77, ff. 91-94.

139 UKNA, PREM 16/1838: Cartledge to Prendergast, 28.04.78.

140 UKNA, FV 14/158: meeting with gen. Bucur on 26.10.78.

141 UKNA, BT 359/22: Mays to Mawson, 01.02.80.

explained to Reginald Ellis, Head of the MoD Sales Department, "[...] as I know you will appreciate, the whole object of the exercise is to produce income for Rolls-Royce. We must sell more of what we have, certainly in hardware and, where it does not harm us in technology. It can be profitable and there is nothing wrong with that".[142]

Rolls-Royce proposed a phased approach that included lengthy preliminary negotiations, the approval of the licence no sooner than 1986 and the delivery of the first engines produced in the United Kingdom in 1988. In this way, it would be possible to make a request of exception to CoCom, highlighting how the Romanian programme was to be initiated about a decade later, when the tactical relevance of the Spey 202 for the military potential of the Warsaw Pact would be minimal. In addition, the transfer of technological know-how provided was very limited compared to what the Romanians had already received with the civil engine manufacturing licence and that would definitely exclude a number of metallurgical techniques considered sensitive. Basically, as remarked by the Department of Industry, the Romanians would find themselves producing the Spey 202 only when its technology – already transferred to the PRC years earlier – had become obsolete. However, even if the assessment of the Defence intelligence concluded that the granting of the licence would not alter in any significant way the balance of power between NATO and the Warsaw Pact, in the atmosphere of the so-called 'second Cold War' both the FCO and the MoD opposed the proposal. The reason given was that the licence for the Spey 202 would still contribute to the development of a Romanian supersonic fighter in the 1990s, allowing it to free up, indirectly, the military-industrial resources of the Warsaw Pact for other aims.[143]

Thus, in early 1981, the prospect of extending Anglo-Romanian cooperation to the military aeronautical sector was definitely shunned. Considering that the Head of the Air Division of the Department of Trade had ruled out that the Spey 202 would pose any insurmountable obstacle at CoCom and that it was generally admitted that the risks from the technical and military point of view were virtually non-existent, it can be understood how Romania's membership of the Warsaw Pact would concretely limit economic and industrial cooperation, regardless of the intolerance towards multilateral controls on the East-West trade.[144]

---

142 UKNA, BT 359/23: Pepper to Ellis, 16.09.80.

143 UKNA, BT 359/23: Wielding to Cresswell, 24.10.80; Mann to Bickel, 6.11.80; Watkins to Sandars, 17.12.80; note by Wilson, 21.1.81; note by Mann, 02.02.81.

144 UKNA, BT 359/21: Maynard to PS/Secretary of Trade, 14.03.78.

The reason for the different attitude towards China was not to be found only in Rolls-Royce's substantial profit expectations: for years the same expectations had driven the British company to keep the question of the Spey 202 unresolved and to insist with the British Government to start negotiations. It was Bucharest's alliance with Moscow, and the hypothetical political repercussions of an agreement, that cooled off the enthusiasm.

These constraints imposed by the membership to the Warsaw Pact were probably further weighed down by the end of Détente and by the marginalisation of Romania on the international scene, where it no longer appeared as a maverick but only as a particularly harsh and abject Communist dictatorship. However, these same constraints were at work even before, when Détente had known its Golden Age, and the Balkan country still aroused the interest Western governments. On the one hand, the invasion of Czechoslovakia and the Brezhnev Doctrine accelerated the Romanian aeronautical projects, but, on the other hand, they put a limit on cooperation with the United Kingdom, which would never be removed. The regime, for its part, would never completely give up its supersonic fighter project, trying unsuccessfully to re-engage Yugoslavia and thinking about employing the MiG-23 engines and then the MiG-29's.[145] Only the fall of Communism and the death of Ceaușescu's would definitively file away the largest, if not the most unrealistic, of Romania's aeronautical ambitions.

145 Petre Opriş, „Avioane de vânătoare Mig-23 şi Mig-29, pentru forţele aeriene române (1977-1989)", *Art-Emis Academy*, 11.03.2012, at http://www.art-emis.ro/analize/875-avioane-de-vanatoare-mig-23-mig-29-pentru-forele-aeriene-romane-1977-1989.html; Costantin Olteanu, „Din relaţiile militare româno-iugoslave (1980-1985)", *Opinia Naţionala*, No. 550 (24.10.2011), p. 1, 4.

CHAPTER 4

Chartering the Modernisation of Civil Aviation

It is no coincidence that in a study published in the UK, in the context of research on the 'Dutch Disease,' the term 'de-industrialisation' was first used.[1] The crisis of the 1970s, along with the difficulties of public finances and rising unemployment, was instrumental in determining the orientation of Anglo-Romanian cooperation in the aviation field after the rapid collapse of the initial prospects. This new phase, however, was still characterised by the competitive dynamics between Western countries, constantly on the lookout for new markets behind the Iron Curtain, and it was also marked by the lure of the political usefulness of transactions with bloc countries. Domestic reasons also played a part in this phase, aiming to support manufacturing capabilities in civil aviation. This was a crucial element in favour of a policy of loans that allowed the sale of five more 1-11s to TAROM.

4.1 Shrinking Perspectives in the UK

In the first half of the 1970s, in Western industrialised countries, it was still considered possible to demonstrate the existence of an inverse relationship between inflation and unemployment. In a given economy, in short, the rise of inflation as a result of expansionary measures would lead to the need for an increased labour force and, consequently, to a decline in unemployment.[2] This idea was essentially a travesty, considering as normal exceptional conditions that had allowed extraordinary economic growth and full employment in the decades following the Second World War. By the late 1960s, however, the situation had changed

---

1 Ajit Singh, "The UK Industry and the World Economy. A Case of De-industrialisation", *Cambridge Journal of Economics*, Vol. I, No. 2 (1977), p. 113-136.

2 Richard Lipsey and William Scarth (eds.), *Inflation and Unemployment: The Evolution of the Phillips Curve*, Cheltenham, Elgar, 2011; Middleton, *Charlatans*, *op. cit.*, p. 288-295.

significantly: on the one hand, the overcapacity of traditional production sectors in Western countries, due to over-investment and enhanced by industrialisation in the Third World, had led to a drop in profitability, especially in manufacturing; on the other hand, a more fragile growth was accompanied by rising inflation, compounded by the ever-increasing cracks in the Bretton Woods system and, at the same time, the difficulty of finding room in the workplace for the baby boom generation.[3]

With hindsight, it is hardly surprising that the consequence of excessive investment in an overheated economy, also in the presence of a saturated market, was stagflation. Equally unsurprising was the failure of traditional Keynesian interventions to stimulate demand. However, for those who lived through those years, it may not have been so easy to understand that the Western world was turning the page, experiencing a difficult and uncertain transition from the Fordist production system to what was later called the post-industrial society.[4] Inflation had been driven to record levels by the oil crisis, thus becoming the most debated economic question of the time, and not only in the United Kingdom. It concerned everyone and, unlike unemployment, not only those who were directly affected.[5] Meanwhile, neo-corporatist approaches based on complex mechanisms of wage bargaining between State, businesses and trade unions continued to enjoy certain popularity during the 1970s.[6]

The transition of the 1970s also represented a crisis in a Schumpeterian sense, that is, the difficult shift to a new technological order. As such, it would mercilessly expose the limitations and the contradictions of planned economies, in which this process failed to materialise. On the contrary, it became very common to invest even more in those industries and technologies that, in the West, were gradually being phased out, especially after the oil crisis.[7] Charles Maier has observed that the developmental crisis of the capitalist world in the 1970s and the terminal crisis of

---

3 Robert Brenner, "The Economics of Global Turbulence", *New Left Review*, I/229 (1998), special issue; Herman van der Wee, *Prosperity and Upheaval. The World Economy, 1945-1980*, Berkeley, University of California Press, 1986, p. 79-198, 454-477.

4 Daniel Bell, *The Coming of Post-Industrial Society. A Venture in Social Forecasting*, New York, Basic Books, 1973.

5 James Alt, *The Politics of Economic Decline: Economic Management and Political Behaviour in Britain since 1964*, Cambridge, Cambridge University Press, 1979, p. 183-201.

6 Philippe Chassaigne, *Les années 1970: fin d'un monde et origine de notre modernité*, Paris, Colin, 2008, p. 25-41.

7 Eichengreen, *The European Economy Since 1945*, *op. cit.*, p. 294-297.

Communism during the 1980s are essentially two phases of the same turbulence in the industrial production system.[8]

Contrary to the situation of the 1930s, the Communist countries were no longer isolated from the rest of the world; indeed, they were bound to it by a considerable and growing network of ties. In this way, the international decline of the old flagship production sectors would inevitably have an even greater impact, mainly because of the structural features of the forced industrialisation of the 1950s and the 1960s, aimed at replicating, in each people's democracy, the same model of extensive growth, based on production with high energy consumption and often characterised by technological obsolescence. In addition, the invasion of Czechoslovakia had, in fact, knocked out a policy of reforms that could, at least in theory, have stimulated a breakthrough in productivity and, therefore, the transition to a different growth model. Having ruled out this option, the ideological change of heart was accompanied by the adoption of an import-led strategy focused on the acquisition of technology from the West to keep pace in terms of productivity and to achieve that much-coveted catch-up with the capitalist world.[9]

It was widely believed that such circumstances would favour an increase in the quantity and quality of East-West economic relations, especially in Europe. In fact, if the Communist countries were in need of Western technology to re-invigorate production processes and maintain high growth rates, stagflation and problems related to the balance of payments in the capitalist world made it even more important to find new markets in the CMEA area. During the first half of the 1970s, Eastern Europe saw an increase of ten percentage points in its share of total exports from the free world, quadrupling imports of machinery and equipment from OECD countries, and doubling technology imports. However, this also entailed an increase in the integration of this area into the global economy, which meant further exposing the planned systems to fluctuations in world market prices during a period of high inflation characterised by an exceptional rise in the price of commodities and intermediate products,

---

8 Charles Maier, "'Malaise'. The Crisis of Capitalism in the 1970s", in Niall Ferguson *et al.* (eds.), *The Shock of the Global. The 1970s in Perspective*, Cambridge (MA), Harvard University Press, 2010, p. 25-48.

9 André Steiner, "The Globalisation Process and the Eastern Bloc Countries in the 1970s and 1980s", *European Review of History*, Vol. 21, No. 2 (2014), p. 165-181; Ivan Berend, *From the Soviet Bloc to the European Union. The Economic and Social Transformation of Central and Eastern Europe since 1973*, Cambridge, Cambridge University Press, 2009, p. 20-26.

which the recently-purchased technologies, themselves a vehicle of inflation, needed in order to be used.[10]

Moreover, the systemic weakness of the planned economies in exploiting these technologies, because of bureaucracy and the inefficiency of quality controls, became immediately evident. Not only did the substantial increase in exports to Western countries that should have marked the success of the import-led strategy for growth fail to materialise – contrary to what was happening with the so-called Asian Tigers – but it was possible to detect the unmistakable signs of a loss of competitiveness in most primary products and many manufactured goods. The hardening of the American position regarding the transfer of technology in the late 1970s, along with the progressive enfeeblement of Détente, marked the collapse of the import-led strategy.[11] Far from achieving the convergence of the two systems, the outcome of these efforts was the increasing marginalisation of the planned economies in the global trends, which heralded a 'globalisation split into two' for the 1980s.[12]

In view of the continuing deficit in the East-West trade balance, the immediate result of this import-led strategy was the surge in foreign debt for the planned economies. Unable to rely on adequate reserves of hard currency, it was mandatory for them to pay for imports through lines of credit with the exporting countries. The real interest rates in a period of high inflation did, after all, justify this course of action, assuming that the loans would be repaid through the sale of what would be produced with the machines and the technology acquired. Unfortunately, not only did these export revenues not materialise, but new debts continued to accumulate thanks to the willingness of Western counterparts – partly to pay off those already incurred in the past, partly because regimes such as the GDR or Poland were forced to ensure an acceptable standard of living for their populations. On the one hand, Western countries were able to recycle the huge surplus of petrodollars deposited with them by the oil producers after the crisis of 1973, and on the other hand, they could seek new markets in the East to face the economic consequences of stagflation. The West,

10 Van Brabant, *Socialist Economic Integration*, *op. cit.*, p. 275; Aldcroft – Morewood, *Economic Change*, *op. cit.*, p. 162-165; Iliana Zloch-Christy, *Debt Problems of Eastern Europe*, Cambridge, Cambridge University Press, 1987, p. 71-72.

11 Friedrich Levcik and Jan Stankovsky, "East-West Economic Relations in the 1970s and 1980s", in Christopher Saunders (ed.), *East-West Trade and Finance in the World Economy. A New Look for the 1980s*, New York, St. Martin's Press, 1985, p. 77-124; Mastanduno, *Economic Containment*, *op. cit.*, p. 199-219.

12 Jürgen Osterhammel, *Globalisation: A Short History*, Princeton, Princeton University Press, 2005, p. 113.

therefore, had an extraordinary opportunity to provide export credit, guaranteeing transactions with the bloc economies, on the assumption that authoritarian forms of government were more reliable, since they did not have to worry about public opinion, and that, if the worst came to the worst, the USSR would step in to settle the debts of its satellites. How wrong these assumptions were is evident from the fact that, in the mid-1970s, Moscow stopped protecting bloc countries from the effects of the oil crisis; these countries witnessed a progressive deterioration in their terms of trade within the CMEA. In this situation, it is easy to imagine the effect that the doubling of the average nominal interest rate of the Eurocurrencies between 1978 and 1981 had on these countries.[13]

Romania also participated fully in these dynamics, albeit with some peculiarities. The directives approved by the Tenth Congress of the RCP provided for the continuation and intensification of the efforts to bridge the gap with the advanced economies, building what was called a 'multilaterally developed socialist society.' This definition served as a reminder that economic cooperation and foreign trade – crucial elements for this development – would not only apply to the Communist countries, but, potentially, to every useful counterpart. In fact, during the 1970s, Romania showed investment rates in fixed capital higher than any other country in the CMEA area, along with a robust growth of the national income. The importance attached by the Tenth Congress to the chemical industry, however, was already a harbinger of disharmonious investments, opposing the trends followed by those advanced economies that the East wanted to catch up with. Similarly, the significant increase in the country's oil-refining capacity was achieved just when its oil reserves were running out and, above all, at a moment when the global demand for the product was on the decline.[14]

As well as untimely investments, Romania shared with the other bloc countries the growth of foreign debt in convertible currency. Its

---

13 Stephen Kotkin, *The Kiss of Debt. The East Bloc Goes Borrowing*, in Ferguson *et al.* (eds.), *The Shock of the Global*, *op. cit.*, p. 80-93; Charles Maier, *Dissolution. The Crisis of Communism and the End of East Germany*, Princeton, Princeton University Press, 1997, p. 59-107; Aldcroft – Morewood, *Economic Change*, *op. cit.*, p. 167-168; Zloch-Christy, *Debt problems*, *op. cit.*, p. 83-87; David Stone, "CMEA International Investment Bank and the Crisis of Developed Socialism", *Journal of Cold War Studies*, Vol. 10, No. 3 (2008), p. 48-77.

14 *Directivele Congresului al X-lea al PCR cu privire la planul de dezvoltare economico-socială a României pe anii 1971-1975 și liniile directoare ale acesti dezvoltari pe perioada 1976-1980*, in *Congresul al X-lea al Partidului Comunist Român*, *op. cit.*, p. 666-716; Aldcroft-Morewood, *Economic Change*, *op. cit.*, p. 169.

role as a maverick, borne out by the decision to join the International Monetary Fund and the World Bank, in spite of the Soviet Union's veto on its satellites' allegiance to international financial institutions, secured preferential treatment for Romania from the West. This also resulted in a greater propensity to grant credit despite the opacity of Romania's economic system. It should be noted, however, that Romania's debt levels remained relatively low compared to those of other Eastern European countries, partly because of the increasingly neo-Stalinist attitude of Ceaușescu's regime. This was epitomised, firstly, in the economic sphere, by the will to resume work, in 1973, on the Danube-Black Sea canal, an infrastructure imposed by Stalin and abandoned soon after his death, and secondly, by Ceaușescu's tendency to gigantism, still partially kept under control by the presence of technocrats. All this meant that hard currency was certainly not used to guarantee acceptable living standards for the Romanian population. Ceaușescu himself insisted strongly on the need to contain the country's debt, saving every penny, while also addressing the necessity to reach an equilibrium in the balance of payments, rationalising import mechanisms (i.e., centralising them) and giving priority to foreign trade, through industrial cooperation with advanced countries and a special focus on developing countries, which would make do with poor quality products while, at the same time, ensuring the supply of raw materials.[15]

Romania's peculiarities – namely, the possibility for the regime to show indifference to the sufferings of the population, Ceaușescu's megalomania and personality cult, the allegiance to international financial institutions, the increasing centralisation in the management of the economy and a strategy clearly aimed at exploiting the fact that Romania was supposedly still a developing country – did not exempt the country from the downward trend of other planned economies. From the mid-1970s onwards, the slowdown in the pace of growth became apparent. Neither the strategy focused on developing countries nor cooperation with the West – for which Romania, first among all members of the CMEA, had legalised joint ventures – gave the desired results. This was essentially because of the constraints of a highly centralised planned economy and the lack of outlets for products, often of poor quality or simply not required. It was in this context that the deficit of the balance of payments in Romania began to

[15] Nicolae Ceaușescu, *România pe drumul construirii societății socialiste multilateral dezvoltate*, Vol. 9, Bucharest, Ed. Politică, 1974, p. 634-637; Ion Bulei, *Brève histoire de la Roumanie*, Bucharest, Meronia, 2005, p. 202-206; Geoffrey Swain and Nigel Swain, *Eastern Europe Since 1945*, Basingstoke, Palgrave, 2009, p. 172-175; Harold James, *International Monetary Cooperation Since Bretton Woods*, Oxford, Oxford University Press, 1996, p. 560-562.

reach alarming levels. Paradoxically, the credit enjoyed by Bucharest thanks to its links with international financial institutions was counterproductive, as it allowed the regime to finance the debt using short-term Euromarket credits, which made the system even more vulnerable. So, in the wake of the 1981 Polish crisis, the loss of the credit lines with Western banks brought about a chain reaction that forced Romania to a first rescheduling agreement in July 1982. At this point, the regime implemented a drastic recovery policy based on exporting as much and importing as little as possible, which resulted in a serious shortage of food and energy for the population. In this way, the government was substantially able to repay the approximately $10 billion debt just before its fall, but at a high cost – the devastation of the country.[16]

In the United Kingdom, the crisis of the 1970s caused a significant shift in emphasis within the declinist paradigm, going from the problem of a growth rate lower than that of other Western economies to the issue of inflation, which reached 25%, as well as currency difficulties.[17] This is reflected in the historiographical interest in the 1976 sterling crisis and the intervention of the IMF, with an emphasis on the pivotal value of this moment to go beyond neo-Keynesian theories and the perception of the United Kingdom as a country increasingly on the decline.[18] Further emphasis was placed on the features of the intervention policies of the Labour Party, which had returned to government in 1974, paying attention both to the industrial side of interventionism and to a neo-corporatist incomes policy.[19]

---

16 Zloch-Christy, *Debt problems*, *op. cit.*, p. 80-82, 89, 114-116.

17 Alec Cairncross, *The British Economy Since 1945. Economic Policy and Performance, 1945-1990*, Oxford, Blackwell, 1992, p. 188-193; James Alt, *The Policy of Economic Decline in the 1970s*, in Lawrence Black, Hugh Pemberton and Pat Thane (eds.), *Reassessing 1970s Britain*, Manchester, Manchester University Press, 2012, p. 25-40; Jim Tomlinson, "British Government and Popular Understanding of Inflation in the mid-1970s", *Economic History Review*, Vol. 67, No. 3 (2014), p. 750-768.

18 See Chris Rogers, "Economic Policy and the Problem of Sterling under Harold Wilson and James Callaghan", *Contemporary British History*, Vol. 25, No. 3 (2011), p. 339-363; Douglas Wass, *Decline to Fall. The Making of British Macro-Economic Policy and the 1976 IMF Crisis*, Oxford, Oxford University Press, 2008; Kevin Hickson, *The IMF Crisis of 1976 and British Politics*, London, Tauris, 2005; Mark Harmon, *The British Labour Government and the 1976 IMF Crisis*, Basingstoke, Palgrave, 1997.

19 Jim Tomlinson, "The Politics of Declinism", in Black – Pemberton – Thane (eds.), *Reassessing 1970s Britain*, *op. cit.*, p. 41-60; Noel Thompson, *Political Economy and the Labour Party*, Abington, Routledge, 2006, p. 189-215; Mark Wickham-Jones, *Economic Strategy and the Labour Party. Politics and Policy-Making, 1970-1983*, Basingstoke, Palgrave, 1996, p. 53-84, 137-157.

Less attention seems to have been paid to the wider impact on economic and techno-industrial cooperation with Eastern countries,[20] which were also considered an important safety valve in the face of the growth crisis in the capitalist West and one of the pillars of the process of Détente. How did the growth of inflation and the credit restrictions affect Britain's ability to conclude transactions with its counterparts behind the Iron Curtain? More importantly, if counter-trade was considered a necessary toll and, at the same time, the main problem to solve in order to find market opportunities in the East, then the issue should be properly analysed, particularly in view of the lack of specific historiographical studies on this topic.[21]

In the case of Anglo-Romanian aeronautical relations, the decision to buy the BAC 1-11s was ultimately motivated by the fact that much of the cost of the aircraft was to be covered precisely by mechanisms of counter-trade and that BAC had offered to Romania the opportunity to cooperate in the production not only of the Islander, but also of the 201, a short-range twin-engine aircraft that would seat up to sixty passengers.[22]

Apparently, the position of the British aeronautical industry in Romania was very solid. With no immediate competitors in the fixed wing sector, it was feasible, banking on the sale of 1-11s, to secure a more solid, well-structured and long-term relationship, based on the clear interest of Romania – a country that for a certain number of years would still have one of the highest growth rates in the bloc area – in reactivating its aviation industry. This interest had been clearly and repeatedly stated. The Romanian authorities had immediately raised the possibility of new business, such as buying planes for Ceauşescu's trips and for the inauguration of the new Otopeni airport near Bucharest. Everything, however, would entail other measures of counter-trade. In addition, on delivery of the first 1-11 on June 14, 1968, the Romanians pointed out that, for them, the transaction represented nothing more than a promising start. In particular, Verdeţ demanded cooperation on the 201 project, whose fate already appeared rather doubtful; for his part, Geoffrey Knight,

---

20 Hugo Radice, "Experiences of East-West Industrial Cooperation: A Case Study of UK Firms in Electronics, Telecommunications and Precision Engineering Industries", in Christopher Saunders (ed.), *East-West Cooperation in Business: Inter-Firm Studies*, Vienna, Springer, 1977, p. 144-151.

21 See Matthias Judt, „Kompensationsgeschäfte der DDR. Instrumente einer europäischen Ost-West-Wirtschaftsintegration?", *Jahrbuch für Wirtschaftsgeschichte*, Vol. 49, No. 2 (2008), p. 117-138.

22 ANR, CC, Chancellery, 1968_22: protocol No. 6 al şedinţei Comitetului Executiv al CC al PCR din ziua de 15.02.68.

Chairman of BAC Weybridge Division, assured the Romanians that BAC was hoping to finalise cooperation and, should the 201 not see the light of day, they would still work on a new project.[23]

Unlike in the military sector, the main obstacle to the realisation of such a rosy prospect did not lie in the effects of the invasion of Czechoslovakia or of the Brezhnev Doctrine, but basically in the intrinsic weakness of the British aircraft industry. However, this was not immediately apparent. The cancellation of the 201 project, in fact, was soon followed by a series of interrelated initiatives. The Romanians would be granted the licence to produce the HS-748, a medium-haul turboprop airliner with an excellent short-field performance that had had great commercial success and whose capacity was similar to that of the defunct 201. It was, in short, a resumption of previous contacts with HSA for a plane designed, thanks to international cooperation, with a country eager to develop its skills in the aviation sector – namely India, which had been given a licence in 1960.[24] Apart from the aircraft, the Romanians would also produce the engine, Rolls-Royce's Dart turboprops, in Romania, starting from the assembly of parts provided by the United Kingdom. Later on, they would use a progressively larger share of components produced in Romania. Related to this project, there was also BAC's offer to build a modern, well-equipped aircraft factory in Romania, and to participate in the design of a wind tunnel for ICPAS. Overall, the package was estimated to have a value of 22 million pounds in 1970.[25]

In addition, BAC had made an offer to the Romanians for cooperation in the production of another plane that was in the pipeline, this time considered critical for the company's industrial future: the 3-11. Conceived as a plane for medium and short distances, but with a cross-section fuselage similar to that of the Boeing 747 or the DC-10, it would offer, both to small regional companies and to the charter market, competitive standards of comfort, together with lower capital and running costs, also thanks to the use of the new RB.211 engine, which Rolls-Royce was developing for Lockheed. The 3-11 was crucial to ensure the company's presence in the civil market and, consequently, employment at the Weybridge and

---

[23] UKNA, FCO 28/247: minute by Randall, 26.02.68; UKNA, FCO, 28/248: Chadwick to Stewart, 20.06.68; record of a meeting between Knight and Verdeţ on 15.06.68; record of a meeting between Knight and Cioară on 15.06.68.

[24] Sören Eriksson, "Newly Industrialised Economies and the Aviation Industry: Economic and Industrial Policy Views in Asia", in Sören Eriksson and Harm-Jan Steenhuis (eds.), *The Global Commercial Aviation Industry*, Abingdon, Routledge, 2015, p. 241.

[25] UKNA, FCO 28/1072: President's visit to Rumania – Steering brief, annex D, n.d.; President's visit to Rumania – Brief 7: aircraft, n.d.

Hurn factories, when the 1-11 ceased to be produced. For this reason, BAC invested significantly in the project and, from the end of 1968, led an aggressive global marketing campaign to put together a list of orders, which would convince the government to authorise the necessary financial aid. The offer to Romania, then, was essentially an element of counter-trade – from which, however, the Romanians would benefit also in terms of technological and productive development – to secure an order for three aircraft.[26]

A year after the delivery of the first 1-11 to TAROM, prospects still looked very promising. The British Ambassador to Bucharest stressed that Romania was interested in increasing the exchanges with the United Kingdom, partly because of the positive experience with BAC, partly because its balance of trade towards France and Germany was already heavily in deficit. Ambassador Pungan was key to this process of re-orientation of Romanian foreign trade in the direction of the United Kingdom. The success of his efforts seemed to be confirmed by the unofficial reassurances of a newly co-opted member of the Executive Committee of the RCP, according to whom the authorities in Bucharest had already taken the decision to conclude the agreement with HSA for the 748 licence. However, on closer scrutiny, dangerous asymmetries emerged between the objectives of the two parties. The Romanians wanted more favourable credit terms – especially concerning duration – and safe outlets for their products on the United Kingdom market, perhaps through a long-term trade agreement. The British, however, seemed convinced that there was no space and no need for special financial incentives for the 748 deal to actually take off. They were also determined not to commit themselves to absorbing a predetermined amount of Romanian products.[27]

Another aspect, perhaps secondary, but certainly troublesome, in Anglo-Romanian relations was the issue of which of the London airports TAROM should use for its service. The airline, which operated a twice-weekly service, had stumbled upon a new air transport policy, which required new operators to use Gatwick rather than the overcrowded Heathrow airport. The Romanians had done their utmost to bring

26 Gardner, *British Aircraft Corporation*, *op. cit.*, p. 178-183; "British Aircraft Corporation Three-Eleven", *Aircraft Engineering*, Vol. 40, No. 11 (1968), p. 22-23; "Three-Eleven Prospects", *Flight International*, 17.09.70, p. 431.

27 UKNA, FCO 28/745: minute by Laskey, 06.06.69; UKNA, FCO 28/760: minute by MacMahon, 10.07.69; Minister of State Office Minute No. 657, 15.07.69; McMeekin to Bache, 21.11.69.

TAROM to the main airport – also for reasons of prestige, as they were the only Eastern bloc country using Gatwick – especially after the acquisition of the 1-11s. However, the British authorities had been adamant, privately admitting that, yes, exceptions had been made in the past, but that Romania was not able to offer advantages or threaten reprisals that could justify such a concession. The problem was destined to resurface because British European Airways (BEA) fully intended to launch twice-weekly flights connecting London to Zagreb and Bucharest, starting from the spring of 1970 and, in an effort to secure passengers, the company would also have TAROM organise package tours that IATA rules did not allow operators to offer on their scheduled flights. BEA operated from Heathrow and it was therefore inevitable that, in view of this cooperation, TAROM would reiterate its demand for admission to the main airport. In addition, the substantial transactions being negotiated at that time (the licence for the 748 and cooperation on the 3-11) made the question all the more sensitive, especially since the Romanians wanted to negotiate a proper air services agreement during Maurer's visit to London, scheduled for autumn 1969.[28]

The importance of aviation problems, and in particular the negotiations on the 748, for the first official trip to the United Kingdom made by a Romanian Prime Minister, was confirmed by the fact that, at the very last moment, Ispaş was included in Maurer's delegation. In terms of diplomatic relations, Ispaş was one of the driving forces behind the scenes in Anglo-Romanian bilateral cooperation, in his capacity as a director in the Ministry of Machine Building Industry. According to the testimony of General Aurel Niculescu, since 1970 Commander of the Romanian Air Force, it was Ispaş himself – a skilled practitioner and a native of Oltenia, like Ceauşescu – that convinced the Secretary General to invest in the revitalisation of the aviation industry.[29] There is, perhaps, an element of exaggeration in this retrospective reconstruction, especially in the light of the role played by scientist Henri Coandă, a founder of modern aerodynamics, who had returned to Romania to direct the National Institute for Scientific and Technical Creation (Institutul Naţional de Creatie ştiinţifică şi tehnica, INCREST), soon to be set up, and to act as

---

28 UKNA, FCO 14/581: Walsh to Sinton, 28.05.69; Use of Gatwick/Heathrow airport – Brief for the visit of the Rumanian PM, 11.69; minute by Giffard, 14.11.69; Sendrea to Gardner, 12.11.69.

29 Sorin Turturică (ed.), *„Pe aviatori, lăsaţi-i să zboare!" Un dialog între general av. (r.) Aurel Niculescu şi Sorin Turturică*, Bucharest, Ed. Anima, 2008, p. 78.

an advisor on issues of science and technology.[30] Up to that point, Ispaş had certainly been closely involved in the negotiations on the 748.[31]

Maurer's visit had the specific purpose of insisting once again on the importance of bilateral economic relations, mutually beneficial for Bucharest's maintenance of a foreign policy that could be independent from Moscow. In order for the benefit to be really mutual, the Anglo-Romanian trade balance had to be redressed, after it had passed from passive to active for the United Kingdom, and, according to Maurer, the most promising area to achieve this was that of technical and scientific cooperation. Wilson confirmed the British interest in reinforcing bilateral relations in this direction, expressing the hope that the air services agreement between the two countries would soon be concluded. He was also in favour of Romania's idea of establishing an aviation working group within the framework of bilateral technological cooperation, given that this seemed to be the sector with the greatest potential.[32]

From the British perspective, these were real concessions, and, since the result of the technological cooperation agreement signed two years before was not considered particularly encouraging, the idea of the aviation working group was not welcomed by MinTech. Also, it was thought more than likely that the air services agreement would offer the opportunity for a new Romanian offensive to move TAROM from Gatwick to Heathrow. However, these concessions did not hit the mark: the Romanians wanted to strengthen their balance of payments, through higher exports paid in hard currency, or with a substantial credit improvement, which was what they were trying to do with their other partners in Western Europe. In the specific case of the 748, the British had no intention of making concessions regarding the financing, both because they believed they had already offered particularly advantageous terms, and because they were convinced that, in the end, for lack of alternatives, the Romanians would accept. The credit line of the Export Credits Guarantee Department (ECGD), in fact, covered 90% of the goods and services of British origin (whereas usually the coverage was of about 80-85%) for a period of five years, provided

30 On Coandă's importance see Hooker, *Not Much of an Engineer*, *op. cit.*, p. 199 and Sergiu Celac's communication to the Author, 17.09.2015. See also Ştefan Bălan and Nicolae Mihăilescu, *Istoria ştiinţei şi tehnicii în România. Date cronologice*, Bucharest, Ed. Academiei RSR, 1983, p. 390.

31 UKNA, FCO 28/744: Bucharest to FCO, tel. 397, 17.11.69; UKNA, FCO 28/747: minute by McCluney, n.d.

32 UKNA, FCO 28/743: Watts to Giffard, 7.11.69; Giffard to Watts, 14.11.69; UKNA, FCO 28/747: record of conversation between the PM and Mr. Maurer at 11.30 a.m. on 26.11.69.

all components were assembled abroad. This was despite the fact that the percentage of components manufactured in the United Kingdom would decrease progressively. Furthermore, an additional loan guaranteed by the ECGD would cover the substantial compensation required by HSA for the sale of the licence.[33] So Wilson's answer to Maurer's call for help as regards the terms of credit was far from auspicious:

> The Prime Minister said that Sir Anthony Percival, the Secretary of the Export Credits Guarantee Department (ECGD), the largest public credit insurance organisation in the world, would be joining the party for lunch. But he must point out that large as ECGD was, its resources were not unlimited. Mr. Maurer commented that everything was limited, but funds more so than other things.[34]

This attitude did not change even when it was revealed that the Romanians had asked Fokker to make an offer for the sale of the production licence for the F.27, a plane that was equipped with the same engine and whose operational functions were equivalent to those of the 748, albeit with a lower load capacity – a licence that had already been granted in the past to an American manufacturer. The emergence of a possible competitor did not necessarily shake the British government's conviction that they had already made the best possible offer. Rather, the attention was focused on rejecting the Romanian attempt to link the Dart to the Viper issue on grounds of a more economical use of resources. Therefore, on his visit to Romania, March 9-11, 1970, President of the Board of Trade Mason was instructed to obtain from his interlocutors the commitment to quickly close the deal with HSA, insisting, on the one hand, that the terms offered were already exceptionally generous and, on the other, rejecting any link between this particular transaction and the Viper or, more generally, the question of any redress in favour of Romania's trade balance.[35]

All of Mason's interlocutors duly brought up the issue of the credit terms, insisting in particular on their duration, which, according to Ispaş, had to be extended from five to seven-ten years. Although HSA seemed to have understood that the transaction was at risk and, therefore, had proposed covering an additional 15% of the cost of the contract in

---

33 UKNA, FCO 28/1072: President's visit to Rumania – brief 6: Anglo-Rumanian Technological Agreement, n.d. and brief 7: aircraft, n.d.; MAEF, série 29, sous-série 29, dos. 269: note sur l'aide à la balance roumaine des paiements, 14.01.71.

34 UKNA, FCO 28/747: record of a meeting held on 26.11.69 at 12 noon.

35 UKNA, FCO 28/1072: President's visit to Rumania – steering brief, n.d.

currency with counter-trade, by producing spare parts in Romania, as well as by putting pressure on the British government to allow an increase in Romanian export quotas, Mason was immovable. Genuinely believing that they had already made a fully competitive offer, Whitehall officials thought it unnecessary, even counterproductive, in terms of competition among Western countries, to return to the problem. This way, however, Mason's position proved negative across the board; he told the Minister of the Machine Building Industry, Ion Avram, that the credit terms would not be changed, and specified to Verdeţ that the trade balance deficit would be structural, at least for a while, as all the products that Romania intended to export – such as butter, bacon, etc. – were already oversupplied globally. To further complete the picture, Mason added that the British government had not yet taken a decision regarding the launch of the BAC 3-11 and, therefore, this prospect of cooperation remained pending.[36]

It was no mystery that the visit had been a dismal failure. Leaving aside the technical aspect, it was politically unsustainable to refuse better credit terms for the 748 contract, and, at the same time, to imply that Romania's credit deficit was irreversible, at least in the short and medium term, especially in view of the fact that, when the trade balance was unfavourable to the United Kingdom, only two years previously, the authorities in London had had no hesitation in underlining the urgency to readjust the balance. Ambassador Laskey, in his report on the visit, remarked that a better understanding of Romania's demands was much needed. The deficit towards Western countries was a given and Bucharest would limit its imports the moment it proved impossible to maintain an acceptable balance, resulting in damage to the British export industries. In May 1970, just as Knight was explaining to the Royal Aeronautical Society how the last years had been amongst the most profitable for the commercial aviation market, contrary to the dire predictions of the Plowden report, and how the pivotal point for sales had become the credit financing of transactions,[37] negotiations on the 748 definitively ran aground. Of course, along with this, all the other industrial actions associated with the 748 came to a grinding halt.[38]

---

36 UKNA, FCO 28/1072: record of a meeting between the President of the BoT and Mr. Burtică on 09.03.70; Walsh to Bellamy, 12.03.70; meeting with Mr. Verdeţ on 11.03.70; UKNA, FCO 14/706: note of a meeting between the President (BoT) and Mr. Avram on 10.03.70.

37 "The Civil Aircraft Industry", *Aircraft Engineering*, Vol. 42, No. 5 (1970), p. 5.

38 UKNA, FCO 28/1072: Laskey to FCO, diplomatic report 233/70, 23.03.70; UKNA, FCO 69/103: Laskey to BoT, tel. 22, 11.06.70.

However, excluding the very positive development of the partial transfer of BEA flights to Gatwick and the company's pooling agreement with TAROM (a development that finally opened Heathrow's doors to the Romanian national airline, while also being a good omen for the negotiation of the air services agreement between the two countries),[39] the evaluation of Anglo-Romanian aeronautical relations was actually very negative, because the prospects for cooperation in the civil sector were suddenly reduced, depending, as they did, on the increasingly uncertain launch of the 3-11.

The unexpected victory of the Conservatives in the elections of June 1970 was not a primary factor in the further worsening of the state of affairs, as the aerospace sector represented – even before the so-called 1972 'U-turn' – an exception to the policy of industrial disengagement followed by the new government, led by Heath. Rather, the shaky state of British public finances, combined with the on-going drainage of funds still incurred by the Concorde project, along with the financial instability of Rolls-Royce, were what determined a dramatic outcome. When, towards the end of 1970, it became apparent that Rolls-Royce was on the brink of insolvency, mainly because of the contract with Lockheed for the development of the RB.211 engine, the government's intervention resulted in the cancellation of the 3-11. The project had already been opposed by the FCO because it was considered a possible obstacle in the negotiations regarding the UK's membership of the European Community, as the 3-11 was a direct competitor of Airbus, in whose realisation the British had been invited to take part. Ultimately, the 3-11 was knocked out by the inability to simultaneously finance, in the civil sector, Concorde, Rolls-Royce and the production of a new airliner.[40]

Rolls-Royce's collapse, therefore, had a large impact on the sector, both at political and industrial level. The Conservatives, champions of the disengagement policy, found themselves going to the rescue of one of the most prestigious companies in the country, after a bankruptcy that had resulted from the mismanagement that the previous Labour government

---

39 UKNA, FCO 14/706: FCO to Bucharest, tel. 107, 25.03.70; Visit of Sir Brimelow to Rumania – Meeting with Mr. Gliga on 11.04.70; Linstead to Broomfield, 26.05.70; UKNA, FCO 14/921: Baker to Russell, 05.07.72; Agreement between the UK and the RSR relating to Civil Air Transport, 19.06.72.

40 Jim Tomlinson, *Government and the Enterprise Since 1900. The Changing Problem of Efficiency*, Oxford, Clarendon, 1994, p. 295-296; Hayward, *Government and British Civil Aerospace, op. cit.*, p. 107-153; *Rolls-Royce and the RB.211 Aero-Engine*, Cmd. 4860, London, HMSO, 1972; "Rolls-Royce Failure", *Flight International*, 11.02.71, p. 181-184.

had failed to detect. Above all, the subsequent cancellation of the 3-11 essentially marked the end of an era of industrial independence in airliner production, leaving BAC in a wretched situation in the civil sector, because the running-down of production of the 1-11 was not counterweighted by any programme for the future. This situation would create about 2,000 potential redundancies in the Commercial Division of the company, as well as annihilating prospects of cooperation with Romania in the aviation field.[41]

4.2 BAC and US Competition in Romania

At the beginning of January 1971, when John Ferguson Smith, managing director of BAC's Weybridge Division, informed Minister Avram of the cancellation of the 3-11, he was amazed by the apparent serenity of his interlocutor, who assured him that he had no wish for the relationship with BAC to be discontinued. The opportunity to inject new life into this cooperation could be the replacement of the TAROM 1-11, which had crashed the previous month, due to heavy fog over Constance airport, with another plane just in time for the tourist season. Now, the fact that the Series 400, which had been sold in 1968, was no longer in production and was not immediately available on the second-hand market, determined a strong possibility, hinted at by the Romanians, that TAROM could replace its fleet of 1-11s with the Series 500, an aircraft that was slightly bigger and driven by a more powerful version of the civil Spey, the MK. 512, through a new counter-trade agreement. Clearly, BAC was willing to consider any possible compensation mechanism, from an extension of the contract for the production of the Islander, beyond what the Romanians were asking, to far more imaginative systems. Finally, Pop proposed the cession of the project for the 201, which Romania could have built together with Yugoslavia with BAC's assistance, then marketing it freely without licence restrictions.[42]

Officials in Whitehall looked with considerable scepticism at the realisation of this prospect, considering the offer of the 201 of dubious value, while also judging excessively onerous the conditions that Romania was planning to impose for the purchase of the six Series 500 1-11s. The

41 Keith Hayward, *The British Aircraft Industry*, Manchester, Manchester University Press, 1989, p. 140-147.

42 UKNA, FCO 28/1532: note by Reed, 08.01.71; Russell to DTI, tel. 7, 29.01.71; minute by Russell, 01.02.71; minute by Russell, 12.02.71.

relatively low importance given to this situation may be understood from a minor episode, namely the fact that no pressure was ever exerted on the new Rolls-Royce, which had been reconstituted through public intervention, so that the company would meet the Romanian demands to comply with the overhauling of the 1-11 motors on the basis of the price agreed in the 1968 contract, in which there was no escalation clause. The contract set a cost of £12,000 per engine, but because of inflation the actual sum had risen to the region of £15,000-45,000, depending on the contingencies of the moment. Basically, although the head of the FCO Trade Policy Department observed that there was an actual risk of once again pushing Romania towards the USSR in the aviation field, it was decided that it was more important to show Rolls-Royce's independence from the government, especially after its rescue, rather than encouraging the company to make sacrifices to place an unlikely order for BAC's benefit.[43]

Actually, since the summer of 1971, attention had focused on BN's growing difficulties and its subsequent collapse, because of the serious consequences for what was at that time the most important feature of Anglo-Romanian aviation cooperation, which had survived the various crises of British industry between 1970 and 1971. While there was no solution for BN on the horizon, the Romanians informed BAC that they had shelved the idea of replacing the TAROM fleet with the Series 500 and that they were going to purchase a second-hand 1-11 from American Airlines. They would later add a second aircraft, to be bought from Philippines Airlines through BAC. As noted by the Commercial Counsellor to Bucharest, Mark Russell, at that point, the revival of Anglo-Romanian cooperation depended on reaching an agreement on the Viper. However, in the meantime, the new Soviet Tu-154 would be made available, with the risk for BAC of being ousted from the Romanian market.[44]

This risk, however, was not so imminent. In line with the policy of buying planes on both sides of the Iron Curtain followed until then, Romania had every intention of initially equipping TAROM with two Tu-154s but, over the course of 1971, a number of technical defects had emerged, in particular regarding the ILS (instrumental landing system) that, contrary to what was claimed by the Western specialised press,[45] did not comply with ICAO standards in the case of CAT II automatic

---

43 UKNA, PREM 15/634: Priddle to Gregson, 30.03.71; Rolls-Royce Difficulties in Romania and Yugoslavia, n.d.; McDonnell to Gregson, 02.04.71.

44 UKNA, FCO 28/1532: Russell to Downey, 09.11.71.

45 "Tupolev Tu-154", *Flight International*, 18.11.71, p. 814-815.

approaches. In 1972, the Soviets informed the Romanian authorities that the necessary changes would be implemented on the Tu-154A, while also stating that they could only provide the original version until the first half of 1974, without being able to say precisely when the new version would actually be available. At that point, the Romanians refused to buy the plane and used the allocated resources to purchase two An-24s.[46] The Tu-154s in their original version were sold to Egypt thanks to extremely favourable contract terms. One of them crashed in July 1974 during a training flight, killing all crew members and pushing the Egyptian government to ask Moscow to take back the aircraft and return the money. Meanwhile, the Soviets started production of the Tu-154A, after making all the necessary technical improvements.[47]

A more dangerous competitor was certainly the United States, who were carrying out a policy of differentiation aimed at rewarding those countries behind the Iron Curtain that implemented internal liberalisation processes or that took an eccentric orbit to Moscow in matters of foreign policy. In this context, Nixon's visit to Bucharest in 1969 had represented a turning point in bilateral relations with Romania and the interest of the administration in the Balkan country was not limited to the general objective of reducing Soviet influence through trade, but rather looked at Bucharest as a possible channel to make it up with China and to mediate a settlement to the Vietnam conflict. Thus, Romania, thanks to Nixon's personal attention, was not a mere dependent variable in the US-Soviet relations, but it enjoyed a privileged position as opposed to other satellites. This was tangibly demonstrated by the special treatment with which the USA implemented unilateral controls on exports.[48]

In the past, TAROM had asked, through the US Embassy in Bucharest, for information on the Boeing 727, but the State Department had not encouraged Boeing to make a tentative offer to the Romanians, so the matter was not pursued. However, the climate had changed and Ambassador Laskey stressed that Ceauşescu's journey to the USA in October 1970 had an economic aim, as Romania wanted to gain the

---

[46] ANR, CC, Chancellery, 1973_5: protocol No. 2 al şedinţei Prezidiului Permanent din ziua de 22.01.73, ff. 2-3; notă No. M242, 15.01.73, ff. 77-79.

[47] "Egypt Buys Russian", *Flight International*, 23.03.72, p. 404; "Tupolev Tu-154/154A", *Flight International*, 24.10.74, p. 577.

[48] Robert Weiner, "The U.S. Policy of Differentiation Toward Romania", in Paul Quinlan (ed.), *The United States and Romania. American-Romanian Relations in the Twentieth Century*, Woodland Hills, A.R.A., 1988, p. 129-145; Joseph Harrington and Bruce Courtney, *Tweaking the Nose of the Russians. Fifty Years of American-Romanian Relations, 1940-1990*, Boulder, East European Monographs, 1991, p. 285-298.

clause of most favoured nation and the credits of Ex-Im Bank. The visit – Laskey noted – had certainly served to make the American business community understand that the administration was far from hostile to the strengthening of relations with Romania. So, while the difficulties of the British aviation industry had repercussions on Anglo-Romanian cooperation, from Washington, Ambassador Corneliu Bogdan wrote that Boeing was interested in selling its 707s to Romania and that, on informing the White House, it had received strong encouragement. This interest would prove very important when Bucharest ran into the usual difficulties (unavailability, uncertainty of delivery times, etc.) upon buying TAROM's long-haul planes from the Soviets, so that, within a couple of years, Romania would buy three 707s, two for the national flag carrier and one for Ceauşescu's trips.[49]

These developments did not escape the well-informed Pop, always intent on devising new ways to boost bilateral aviation cooperation that, in the civil sector, in early 1972, seemed to have no prospects for the future. Pop knew the plans for modernisation of the TAROM fleet to a tee and, along with Russell, he could not but rejoice that the Romanians intended to buy three Il-62s to cover the long-haul routes: much better the Soviets than the Americans, especially because they would certainly have used the first transaction to try and completely undermine BAC, replacing the 1-11s with their planes. For the moment, though, with the problem of BN's collapse still unsolved, it was unthinkable to raise the offer of the Series 500. All the same, it was necessary to set a fixed deadline of one year – that is, before the Tu-154 became available – to try and give events a favourable turn, playing on Bucharest's desire to buy another second-hand 1-11 and, more generally, to cultivate relations with some key figures in the aviation industry taking advantage of the Viper negotiations.[50]

Indeed, the JUROM dossier represented a suitable framework to reinforce these relations and to create an atmosphere of fraternisation. Coandă, for example, was made an honorary member of the Royal Aeronautical Society on the occasion of a visit to Derby.[51] In mid-1972, he also accompanied Bodnăraş on his private trip to the UK,

---

[49] UKNA, FCO 28/1054: Laskey to Giffard, 09.11.70; ANR, CC, Foreign Relations, 1971_121: notă adresată de MAE lui tov. Ceauşescu, No. 80418, 27.04.71, ff. 37-38; Petre Opriş, „Tatonări ale regimului de la Bucureşti privind achiziţia unor avioane ‚Boeing', 1964-1974", *Arhivele Totalitarismului*, Vol. XXII, No. 82-83 (2014), 81-92; *Id.*, „Avioane civile utilizate de şefii statului român (1957-1991)", *Romanian Journal of History and International Studies*, Vol. 1, No. 2 (2014), p. 241-254.

[50] UKNA, FCO 14/1046: Russell to Watts, 25.01.72; Russell to Watts, 22.02.72.

[51] "Henry Coanda Visits Britain", *Flight International*, 28.01.71, p. 115.

at the conclusion of which the Vice President of the Council of State praised the sincere and reliable British against the 'devious and slippery French.'[52] Until his death, on November 25, Coandă served as a source of information for Pop and the Embassy in Bucharest. Sometimes, such fraternisation could become even excessive and embarrassing, as when Coandă described to Ambassador Ashe the bizarre idea of creating a fleet of hovercraft equipped with tactical nuclear weapons through trilateral cooperation between Romania, the United Kingdom and France (to whose arms industry, according to British intelligence, the ageing scientist had remained tied until the end of the 1960s) to somehow balance the Soviet fleet in the Black Sea.[53]

The same was true for Ispaş, himself a member of the Royal Aeronautical Society from 1973, who had played key roles in the techno-structure of the Romanian procurement and whose influence – before he turned to an academic career, leading to a professorship at the Polytechnic Institute of Bucharest and his 1981 appointment as Rector of the Military-Technical Academy[54] – was reinforced during the reorganisation of the aviation industry that took place in 1972. In fact, in order to increase efficiency and safety, the Civil Aviation Council, under the Ministry of Transport, was replaced by the TAROM-Civil Aviation Command as the sector's single regulatory body under the Ministry of Defence.[55] In fact, the military had long performed many basic technical roles in civil aviation, but the reorganisation was important because it excluded from the chain of command General Laurian Medvedovici, head of the now-defunct Civil Aviation Council. He had worked for the modernisation of the sector and for the establishment of a special school for personnel training.[56] According to Pop, Medvedovici argued that this modernisation was to be achieved with Soviet aircraft and he was against the idea of buying the Series 500 1-11s, which would have led to growing tension with Ispaş, who was far more favourable to this idea.[57]

---

52 UKNA, PREM 15/1185: Ashe to Bullard, 04.07.72.

53 UKNA, FCO 69/294: Ashe to FCO, tel. 187, 01.06.72.

54 Ispaş's bio is available at http://aman.ro/betawp/wp-content/uploads/personalitati/I/ispas%20stefan.pdf.

55 Decretul No. 419/1972 privind trecerea activităţii aeronautice civile şi a unităţilor de aviaţie civilă la Ministerul Forţelor Armate în vigoare de la 01.11.1972, available at http://lege5.ro/.

56 Manuela Vârlan and Doru Vârlan, *Şcoala superioară de aeronautică civilă*, Asociaţia aerospaţială română, 2010, p. 24-31.

57 UKNA, FCO 14/1046: Russell to Cochlin, 23.11.72.

In July 1972, Pop had suggested to Bodnăraş that Romania should create a charter company to meet the demands of the 1973 tourist season by purchasing five to seven Series 500 1-11s for about 20 million pounds. Considering that the increase in TAROM's traffic volume was down to its international charter flights, which in terms of passenger numbers had already surpassed the full service airlines, unsurprisingly the Vice President of the State Council urged BAC to prepare a draft contract, implying that the transaction would soon be completed.[58] In the autumn of 1972, BN's bankruptcy having been sorted out thanks to the purchase of the company by Fairey and having concluded the Viper licence agreement, the prospect looked a lot more promising. Bilateral relations were back on a positive footing, as shown both by Bodnăraş's interest and Ispaş's enthusiasm for closer Anglo-Romanian aerospace cooperation which could be extended from the hard core of the Viper engine licence to BAC and HSA for airframes.[59]

Unfortunately, the cancellation of the contract for the three Soviet Il-62s in October 1972 meant that in all likelihood Boeing would be able to sell its 707s. At the beginning of 1973, Commercial Counsellor Russell noted with alarm that this possibility was extremely dangerous, because Boeing was interested in venturing into the short-to-medium-haul aircraft sector. With American financial support and counter-trade, there was a real risk that, once the company got its foot in the door, it would gradually squeeze out the British from the Romanian market.[60] Indeed, the financial package offered by Boeing for the 707s was exceptional in terms of competitiveness – with two ten-year loans of 20 million dollars each by Ex-Im Bank and Morgan Guaranty Trust, leaving just 10% of the value as a cash outlay. Boeing's purpose, strongly supported by the White House and the State Department, was precisely to dethrone BAC with its 727, even if it meant taking on the seven TAROM 1-11s, thus guaranteeing to the Romanians the homogeneity of the fleet and the resulting economies of scale. Significantly, immediately after the signing of the contract for the 707s, the CIA produced a study entitled *Romania: A New Market for US High-Performance Civil Aircraft*, in which they noted the importance of commercial penetration of Romania to break the Soviet stranglehold on Eastern European markets.[61]

---

58 UKNA, FCO 28/2008: Reed to Burns, 25.07.72.

59 UKNA, FCO 28/2010: Russell to Green, 22.12.72.

60 UKNA, FCO 28/2346: Russell to Green, 23.01.73.

61 NARA, ADD, RG 59: Bucharest to SecState, tel. 856, 09.03.73; Bucharest to SecState, tel. 2966, 15.08.73; CIA Electronic Reading Room, Romania: A New Market for US High-Performance Civil Aircraft, May 1973, CIA-RDP79T01098A000100070001-8.

At a meeting with the President of Boeing, Ernest Boullioun, on March 8, 1973, Ceauşescu insisted that the 707 transaction would serve to develop cooperation for the construction in Romania of a short-haul 60-seater plane. All this aroused Boullioun's interest, although he was aware of the difficulties in achieving that kind of cooperation. He explained that part of the assembly might be subcontracted and, subsequently, a draft project could be defined. If suitable market outlets were found, the project could be launched towards the end of the 1970s.[62] This answer may not have satisfied Ceauşescu, but in fact Romania was placed in a very advantageous position, not only because Boeing was one of an increasing number of companies that wanted Romania to have a special status as most favourite nation, but also because Boeing, Douglas and BAC had begun to vie with one another as to who was able to make the best offer for planes to expand the Romanian fleet. However, BAC started with a handicap because the production line of the Series 500 1-11 had been shut down for lack of orders. Reactivating it would entail an extra cost of about £2.5 million that the company wanted the British government to contribute to. Peter Walker, Secretary of State for Trade and Industry, strongly supported BAC's request.[63]

The oil crisis of 1973 seemed to slow things down. The United States Embassy in Bucharest blamed this on the uncertainties generated by the new international situation. The crisis, in fact, hit the air transport sector both directly (with increased fuel costs) and indirectly (with the drop in traffic volume corresponding to the general economic downturn) at a time when companies were recovering from cash-flow problems and overcapacity in the 1960s. The result was a drastic downsizing of the civil aviation market and the increased importance of efficiency in consumption.[64]

Indeed, the Romanians did give the matter some thought. The effort to improve their performance in civil aviation had continued with new interventions on the organisational structure,[65] but even more so by politicisation, with RCP members occupying leadership roles and positions

---

[62] ANR, CC, Foreign Relations, 1973_21: notă de convorbire No. 745, 15.03.73, ff. 2-10.

[63] UKNA, FCO 28/2347: FCO to Bucharest, tel. 80, 26.10.73; Ashe to DTI, CREDA 80, 29.10.73; UKNA, BT 359/8: Ashe to DTI, CREDA 57, 29.08.73; NARA, ADD, RG 59: Bucharest to SecState, tel. 3993, 01.11.73; Harrington – Courtney, *Tweaking the Nose of the Russians*, *op. cit.*, p. 348-352.

[64] NARA, ADD, RG 59: Bucharest to USDoC, tel. 100, 10.01.74; Bucharest to SecState, tel. 7700, 12.02.74; Hayward, *The British Aircraft Industry*, *op. cit.*, p. 159-164.

[65] Decretul No. 162/1973 privind stabilirea normelor unitare de structura pentru unitatile economice în vigoare de la 11.05.73, available at http://lege5.ro/.

as airline personnel. By the end of 1974, there were over 2,300 party members in the civil aviation and all staff had to go through the political and ideological training system. The results, however, were poor. Wastage and inefficiency persisted, safety measures were inadequate and foreign passengers complained about delays, poor hygiene and, more generally, about treatment that did not match up to international standards. On two occasions, in Zurich and Paris, the local police had to intervene to protect TAROM staff from passenger rage.[66]

In the period 1971-74, foreign tourists to Romania – mainly from the FRG – had generated $360 million in hard currency for the state coffers, with a strong annual growth rate of revenue. Considering that the three 707s, that is, the largest and most expensive aircraft, had cost $45 million and that about half of the TAROM fleet did not conform to Western standards, at first glance the determination to go ahead with the modernisation of the fleet would seem to make economic sense. In fact, the increase in revenue was mainly due to the raising of tourist tariffs (the average amount per person had almost doubled between 1970 and 1974 and a further increase of 20% was envisaged for 1975), irrespective of the services offered, while the number of visitors had already begun to decline in 1973.[67]

If, therefore, the decision to proceed with the purchase of new aircraft even after the oil crisis may have seemed paradoxical, the period that, according to the United States Embassy in Bucharest, marked a break in the negotiations, also marked the fact that BAC had consolidated its package of proposals. It should be noted that the Romanians had promised the British negotiators that they would be confidentially informed if the Americans offered more advantageous conditions and that nothing binding would be signed before a counter-proposal had been received.[68] In reality, Bucharest unscrupulously adopted a policy of presenting each contender with the proposals received from the others, pressuring all of them to make their lowest possible offer. This was particularly so in the case of Douglas, who seemed to have been pretty much out of the running from autumn 1974, when Ceauşescu told the company's Vice President

66 ANR, CC, Chancellery, 1974_112: notă No. S/M.5455, 27.09.74, ff. 15-22; notă priveşte unele deficienţe în activitatea de transporturi aeriene civile, No. 388, 17.09.74, ff. 24-28.

67 ANR, CC, Chancellery, 1974_112: notă privind activitatea de turism internaţional ce se realizeasă organizat prin agenţiile de voiaj străine, pe relaţia Vest, No. 4504, 12.10.74, ff. 29-49.

68 UKNA, FCO 28/2347: Russell to Cochlin, 02.11.73.

that their offers were the least favourable. Then, just a few days before the contract was due to be signed, Douglas was asked to make a new offer, simply for the purpose of putting pressure on BAC.[69] However, even in the case of Boeing, it is significant that the negotiations were resumed when the British offer had already been set up.[70]

From the point of view of counter-trade, in addition to the continued production of the Islander – whose terms the Romanians would have to negotiate directly with the new BN – BAC's initial offer entailed that Romania would manufacture some of the 1-11 components, which would then be bought by the company, allowing the Romanians credit corresponding to the difference in production costs compared to the United Kingdom. To avoid complications with CoCom, the work packages limited the manufacture of details from raw materials to a minimum and they completely excluded the systems, concentrating instead on the structural mechanics. Overall, this would cover about 20% of the capital cost of the planes and spare parts. Furthermore, Rolls-Royce's contribution would be added in the form of a subcontracting agreement for some metal components of the RB.211. The offer, however, was considered to be insufficient and was therefore radically modified. Rolls-Royce's subcontracting arrangement was redrafted so as to guarantee more cash inflow in the first years following the signing of the contract and three 1-11s were lent to TAROM free of charge to cover the tourist seasons of 1975 and 1976. More importantly, the Romanians were invited to cooperate in the production of a new batch of Series 475 1-11s (a plane with the same fuselage as the Series 400 but with the wings and engines of the Series 500 and intended to compete with the Fokker F.28). This would involve a significant transfer of know-how and the possibility for the Romanians to send their engineers to train at the BAC plants.[71]

This cooperation plan, known as ROMBAC, did not seem to convince the Romanians at first, mainly because the concrete prospects depended on the rather remote possibility of further developments of the 1-11, which – in the light of recent experience – the Romanians were dubious about. So, instead of clinching the deal as BAC had hoped, the offer prompted

---

69 ANR, CC, Foreign Relations, 1974_266: stenogramă convorbirii tov. N. Ceaușescu cu vicepreședintele executiv al firmei McDonnell Douglas Aircraft Corp., 31.10.74, ff. 1-8; NARA, ADD, RG 59: Bucharest to SecState, tel. 5042, 11.11.74; Bucharest to SecState, tel. 1519, 09.04.75.

70 NARA, ADD, RG 59: Bucharest to USDoC, tel. 3293, 30.07.74.

71 UKNA, BT 359/8: Brooks to Haynes, 29.01.74; minute by Hadley, 15.02.74; memorandum by BAC, 11.02.74; BAC cooperation offer, 25.06.74.

the Romanians to reopen contacts with Boeing. They offered Boeing the opportunity to secure the TAROM order, pointing out, however, that BAC's offer was more appealing in terms of financing and counter-trade.[72] The financial aspect consisted in securing a loan covering 90% of the cost of the aircraft from Lloyds Bank to the Romanian National Bank for Foreign Trade, to be repaid in forty instalments over ten years, starting from the date of delivery of each aircraft. This would allow a preferential interest rate of 7%, that is, less than half the level reached by inflation in the United Kingdom.[73] A further advantage, not mentioned to the Americans by the Romanians, was the fact that BAC had agreed to improve contractual guarantees on the planes in view of the repeated and often embarrassing failures that TAROM's 1-11s were subject to, mainly concerning air-conditioning and cabin pressurisation, for example when the official delegations of Egypt, Jordan and Mexico were grounded on their visit to Bucharest for the thirtieth anniversary of the fall of Marshal Ion Antonescu's regime.[74]

The success of BAC in the last months of 1974 depended above all – although not only – on the ability to deal with some financial issues and, ultimately, on relevant domestic policy choices by the British government. The deteriorating economic situation in the UK and, in particular, the worrying growth of inflation, had pushed BAC to include an escalation clause tied to the price of the aircraft, since the company was not able to predict the increase in costs accurately. The Romanians were strongly averse to the escalation clause, and the situation was further complicated by the return of the Labour government after the General Election of February 1974 and the implementation of their programme to nationalise aerospace and shipbuilding. This nationalisation soon became a political battleground and resulted in a long dispute over compensation.[75]

In such a situation, BAC asked the government to cover the costs of reactivating the production line of the Series 500 1-11 in order to avoid

---

72 UKNA, BT 359/8: Dorman to Gingell, 28.06.74; NARA, ADD, RG 59: Bucharest to SecState, tel. 3494, 10.08.74.

73 UKNA, BT 359/8: BAC cooperation offer, 25.06.74. See Jim O'Donoghue, Louise Goulding and Grahame Allen, "Consumer Price Inflation Since 1750", *Economic Trends*, No. 604 (2004), p. 38-46.

74 ANR, CC, Chancellery, 1974_112: notă privește unele deficiențe în activitatea de transporturi aeriene civile, No. 388, 17.09.74, ff. 24-28; ANR, CC, Economic Section, 1975_45: notă privind unele deficiențe de comportare în exploatare a avionelor BAC 1-11-424 și prevederi incluse în contractul de import a avionelor BAC 1-11-525, 28.04.75, ff. 2-4.

75 Hayward, *The British Aircraft Industry*, *op. cit.*, p. 148-153.

the risk of the financial outlay being deducted from the compensation due to shareholders. To ensure that the operation could make a profit, it was necessary to produce at least ten planes, but there were only the five for TAROM, and therefore an underwriting was required that would cover a batch of five complete aircraft, a second batch of five sets of details, cost overruns and interest on capital. According to Ferguson Smith, if this operation failed to materialise, the Civil Aircraft Division's activities would be radically reduced, thus favouring Boeing's efforts to oust the British completely from the subsonic aviation market.[76]

Such an operation, in the eyes of Whitehall officials, posed a problem of principle: it was a matter of helping a perfectly solvent company before the need to do so arose, guaranteeing unquantified losses that had not yet occurred. This did not take into account that BAC's shareholders – GEC and Vickers – were asking the government for a formal guarantee that the reopening of the production line would not affect compensation, before the mechanisms for its implementation were put in place. Nevertheless, Benn – the Secretary of State for Industry – interposed in favour of the underwriting, emphasising that after HSA's deferral of the project for the new short-haul airliner (the HS.146), due to its poor market prospects, it was necessary to keep the 1-11 alive in order to lend credibility to BAe's future in the civil sector and, above all, to avoid a further increase in unemployment in the Hurn area, where the rate was already above the national average. So the Treasury agreed to the underwriting, under certain conditions, but it flatly refused to concede any guarantee to shareholders.[77]

Thanks to this help, BAC was able to offer the planes at a fixed price and an agreement was signed on December 30, 1974. Significantly, it was the British trade attaché to Bucharest that informed the American Embassy, while the Romanians hastened to explain to the American diplomats that it was a 'convention,' which carried no binding commitment to actually buy from BAC. While Boeing and Douglas, to varying degrees, allowed themselves to be manipulated by the Romanians, even though they knew that they now stood very little chance of beating BAC, the ROMBAC project was more precisely and more organically defined. It was based on a consignment of twenty planes, with nine distinct packages, focusing on traditional structural mechanics but providing detailed manufacturing

76 UKNA, T 225/4095: Jardine to Cory, 15.11.74; record of a meeting with representatives of BAC, 14.11.74.

77 UKNA, T 225/4095: meeting wth DOI, on 29.11.74; Benn to Barnett, 17.12.74; Barnett to Benn, 20.12.74.

elements in five of them. Systems were not included, materials would be supplied directly by BAC with CAA certification, and the Romanians would use their own machine tools. In 1976, very slowly and with constant delays, the new production line was set up at the IRMA facility, manned by top-quality fitters who had learnt their skills while working on the Islander. Separately, in January 1975, a Memorandum of Understanding was signed with Rolls-Royce regarding its part in counter-trade.[78]

The final act of the negotiations was an example of 'double' manipulation. While the Romanians used the Americans to put pressure on BAC, the company's shareholders took advantage of the spectre of Boeing to ask again for a guarantee that the investment would not result in a cut in the compensation due to them because of nationalisation. In essence, without the guarantee, the Government could choose whether to finance the entire transaction (including the working capital) or see the negotiation fail. Although the policy of the Treasury was not to give in to what was basically blackmail, the prospect of facing the unions and having to commit even more resources at a time when the state of public finances was not exactly rosy meant that in the end the guarantee was granted.[79]

In the following years the underwriting agreement became the subject of controversy, both because of the effects of inflation on costs (the planes, with an initial estimated cost of 2.5 million pounds each, were delivered between January and August at a cost of 3.3 million each), and because the extension of the 1976 agreement and the complex system of the sharing of profits meant that by the 1980s it was no longer clear who owed what to whom.[80] However, on March 24, 1975, the sales contract was signed. Seven years after the 1968 exploit and in much more difficult circumstances, BAC managed to hold its position on the Romanian market, thus deflecting, at least in the short term, the pressures to opt for a future as a sub-contractor for the United States aviation industry or to participate in the European collaboration proposed by the Spinelli

78 UKNA, BT 359/8: Rogers to Gingell, 20.02.75; Rombac 19.4, 13.02.75; memorandum of agreement, 16.01.75; DR 8/27: Rombac 19.2, January 1975; UKNA, DR 8/31: Romania Islander Production Line, report No. 11, 17-21.11.75; Romania Islander Production Line, report No. 13, 25-29.10.76.

79 UKNA, T 225/4095: minute by Cory, 15.01.75; Benn to Barnett, 21.01.75; minute by Scholar, 04.02.75; Barnett to Benn, 05.02.75.

80 UKNA, FV 14/86: minute by Jardine, 05.01.77; UKNA, FV 14/158: minute by Wilde, 10.10.78; UKNA, FV 14/86: minute by Jardine, 05.01.77; UKNA, T 225/4096: Jardine to Cory, 03.04.75.

report.[81] The relaunch of Anglo-Romanian aeronautical relations did not only allow the creation of LAR (Liniile aeriene române), the charter division of TAROM, but it also saw the ROMBAC project evolve from an element of counter-trade into the single most ambitious collaborative venture in aviation between the two countries.

81 "Action Programme for European Aerospace", *Flight International*, 16.10.75, p. 553. See Keith Hayward, "European Union Policy and the European Aerospace Industry", *Journal of European Public Policy*, Vol. 1, No. 3 (1994), p. 347-365.

Chapter 5

Rombac

While the negotiations for obtaining the licence to manufacture the BAC 1-11 in its entirety were undoubtedly the culmination of over ten years of Anglo-Romanian aeronautical relations, these talks also shed light on their inherent structural limitations, especially now that the Romanians were no longer in a position to play their usual card of generating competition between the European contenders. Moreover, both the United Kingdom and Romania were about to enter a new phase coming in the aftermath of the 1970s crisis. The ROMBAC project thus came to epitomise the fact that London no longer needed to maintain a relationship with Romania for foreign policy reasons, while also symbolising the megalomaniacal stagnation of the Bucharest regime. So, just when Anglo-Romanian aeronautical relations seemed to be making breakthrough, they were actually petering out.

5.1 Competing with the Germans

Towards the end of the 1970s, a study carried out on behalf of the World Bank highlighted that Romania had made huge progress over twenty-five years thanks to a catch-up strategy based on high levels of saving and investment. Over the 1950-75 period, by mobilising resources and limiting consumption, Romania had had the highest growth rate in Eastern Europe, sustained by a phenomenal amount of investments (with an annual average growth rate of 13.1%), most of which was ploughed back into industrial production. Thanks to enormous efforts, this development was reflected in the increased exportation of manufactured goods, with some success stories, such as the very popular Romanian farm tractors. However, despite the attempts to acquire modern production technologies and stimulate efficiency and quality through foreign trade, commercial exchanges with the more advanced Western countries were still largely dominated by unsophisticated products, while the deficit was financed through credits.[1]

Although the first half of the 1970s represented a bona fide Golden Age when it came to opening up exchanges and, in particular, for the importation of embodied technology, it soon became apparent that this was not bringing about any significant reduction in the East-West divide in terms of jobs and capital. In 1975-76, in the wake of the oil shock and the slowdown of the global economy, Eastern European countries, with the sole exception of the USSR, tried to remedy the growing trade deficit with the advanced economies and the chronic lack of hard currency, on the one hand, by reducing imports of capital goods from these countries and, on the other hand, by placing greater emphasis on industrial cooperation. Romania also participated in these recovery efforts, although, according to OECD data, it remained only peripherally involved in the East-West industrial cooperation.[2]

The same trend was reflected in Anglo-Romanian trade relations, in the sense that Bucharest succeeded first in absorbing the trade deficit, which had given rise to bitter complaints, while then recording, though briefly, a surplus in the first half of 1976.[3] The point is that this fleeting success coincided with a substantial increase in commercial exchange, which in subsequent years turned out to be far more beneficial to the United Kingdom than to its counterparts behind the Iron Curtain. Nevertheless, the aviation sector remained, at least partially, an exception to the general rule. Indeed, not only did the actuation of the credit line for the purchase of the 1-11s happen at the very moment when Romania was trying to limit imports from Western countries,[4] but more importantly, the industrial cooperation with BAC took a new direction with the granting of the licence to construct the complete aircraft. Thus, the aviation sector stood as a significant exception to Romania's relatively low participation in East-West industrial cooperation, going on to become an onerous and disproportionate commitment for a creaking economy (in 1978 growth of gross industrial output fell to below 10% for the first time since 1956 and,

---

1 Andreas Tsantis and Roy Pepper, *Romania: The Industrialization of an Agrarian Economy Under Socialist Planning*, Washington D.C., The World Bank, 1979, p. 1-14, 202-216.

2 Helgard Wiener and John Slater, *East-West Technology Transfer. The Trade and Economic Aspects*, Paris, OECD, 1986, p. 18-24, 306.

3 UKNA, FCO 28/3080: Prendergast to Cartledge, 20.10.77.

4 Lloyds Banking Group Archives (Lloyds), London, HO/D/BOA/AGE/45/9202: index to advances, Board of 21.11.75; Lloyds, box 5786: minutes of a meeting of the Directors held on 20.08.76.

even more significantly, in 1979 investments increased by 'only' 5.1%).[5] This positive trend came about thanks to the surprising continuity of the policies of the five-year plans of 1971-1975 and 1976-1980, despite the fact that the end of the Glorious Thirty had marked the transition between the two.

The plan for 1976-80 declared a commitment to the continuation and completion, through a 'techno-scientific revolution',[6] of what had been initiated in terms of industrial growth during the previous five years. The aim was to enable Romania, as a developing country, to overcome the economic stage of its growth, according to the ideological paradigm of a 'multilateral developed socialist society', as a step towards the realisation of communism.[7] So, the RCP's plan envisaged that 'greater attention would be paid to the development of air transport',[8] which was supposed to increase by 50% over the 1971-75 period by building up TAROM's fleet with modern planes and increasing commercial aviation activities.[9]

Practically speaking, this would mean not only buying new aircraft, but above all, setting up a national production capacity in line with the discussions held in February 1968 concerning the relaunch of the Romanian aeronautical industry.[10] The interest in the BAC projects, plus the intention to purchase the licence to manufacture the 748 and the contacts with Fokker regarding the F.27 were all attempts to enable Romania to build a short-haul aircraft that could be used both for passenger and freight transport or even – as a customised version – for military purposes. While in the early 1970s the prospects for Anglo-Romanian cooperation had gradually fizzled out, contacts with Fokker had gone ahead. The Dutch company was the third most complete civil airframe industry in Europe, with many successes to its name, but

---

5 Marvin Jackson, *Romania's Economy at the End of the 1970s: Turning the Corner on Intensive Development*, in *East European Economic Assessment. A Compendium of Papers Submitted to the Joint Economic Committee, Congress of the United States*, part 1: *Country Studies, 1980*, Washington D.C., U.S. Government Printing Office, 1981, p. 232-298.

6 Nicolae Ceaușescu, *Dezvoltarea și modernizarea bazei tehnice-materiale a socialismului în România*, Bucharest, Ed. Politică, 1978, p. 79-85.

7 *Id.*, *Raport la cel de-al XI-lea Congres al Partidului Comunist Român*, Bucharest, Ed. Politică, 1974, p. 47-53.

8 *Programul Partidului Comunist Român de făurire a societății socialiste multilateral dezvoltate și înaintare a României spre comunism*, Bucharest, Ed. Politică, 1975, p. 73.

9 *Directivele Congresului al XI-lea al PCR cu privire la planul cincinal 1976-1980 și liniile directoare ale dezvoltării economico-sociale a României pentru perioada 1981-1990*, Bucharest, Ed. Politică, 1974, p. 43.

10 ANR, CC, Chancellery, 1968_29: notă cu privire la studiile preliminare privind stabilirea Industria Aeronautică Română, n.d., ff. 38-59.

because of its size, it needed to think about widening its horizons in terms of international cooperation. This is why it merged with Vereinigte Flugtechnische Werke (VFW), a group that included the old Focke-Wulf and other companies active in the Bremen area. The VFW-Fokker joint venture, seen as a possible forerunner of a genuine rationalisation of the European aeronautical industry and also as an alternative to the consortium model, somehow inherited Germany's first civil aircraft project since the Second World War, the VFW 614. An industrial cooperation agreement for the project was set up with Romania, but it came to an end when the cost of the 'little German Concorde' brought about VFW-Fokker's acrimonious liquidation.[11]

On paper, the 614 was an excellent machine and a perfect choice for the Romanians. A low-wing aircraft specifically designed for economical short-haul operations on low density routes, it offered good cruising speed and climb performance and its wing aerofoils meant that it was suitable for use on short runways. It was powered by two M45H turbofan engines, produced jointly by Rolls-Royce and SNECMA. These were certainly quieter and more up-to-date than the Spey engines and they were able to withstand the stress of take-off and ascent at a significantly lower maintenance cost. Unfortunately, the development plan was overoptimistic and the crash of the prototype in February 1972 certainly did not help. The collapse of Rolls Royce and SNECMA's withdrawal were a setback to the development of the engine and made it much costlier than foreseen.[12]

In the face of worrying financial problems and greatly reduced sales prospects, an agreement with Romania might have seemed useful in order to reach markets that were normally inaccessible to the West. It could have lent political clout to the project in the context of Ostpolitik and thus raised hopes of getting more financial support from Bonn. The Memorandum of Understanding on the 614 project was finally signed on the occasion of Ceauşescu's visit to the FRG on June 26-30, 1973, the first visit to Western Germany by a leader from a Warsaw Pact country. Under the agreement, Romania was licensed to manufacture one hundred aircraft (fifty for domestic use and fifty for export) over a period of 13-15 years. Also in this case, the Romanians would be

[11] Keith Hayward, *International Collaboration in Civil Aerospace*, London, Pinter, 1986, p. 107-110.

[12] "VFW 614 – The Quietest Jet Transport in the World", *Air Engineering*, Vol. 46, No. 6 (1974), p. 5-34; "VFW Crash Investigation", *Flight International*, 10.02.72, p. 207-208.

responsible for producing some of the structural components, but they would have to purchase the systems from or through VFW-Fokker and they would not receive the know-how necessary to develop the aircraft, not even in the form of data relating to flight certification tests. In addition, the agreement bound TAROM to the purchase of two 614s. In order to realise the project, however, a financial outlay of up to 350 million marks was required and the Bonn government was asked to pay a sales subsidy of around 1.5 million marks per plane. The financial situation, therefore, was problematic and negotiations dragged on, with Romania repeatedly asking the FRG for soft loans.[13]

The credits issue caused strain between Bucharest and Bonn, because the German government was not prepared to apply the same conditions to Romania as those for developing countries and, in the difficult global situation, it had no intention of offering public guarantees for private commercial financing of projects like the joint venture between VFW-Fokker and the Romanians.[14] Conversely, after the further purchase of the 1-11s from BAC, Anglo-Romanian relations were flourishing; when Ştefan Andrei, Secretary of the RCP, was sent by Ceauşescu to invite Wilson to visit Romania by the end of 1975, the Prime Minister recalled that purchase as a key moment in furthering bilateral economic understanding.[15]

For his part, Wilson seemed favourably disposed towards Romania, well beyond what might be expected if his general objective was merely to encourage the country to pursue an independent foreign policy. This could have been down to the fact that, in the past, the Romanians had seemed rather distrustful of Wilson and would have preferred the Conservatives to remain in power – a view spread by Pop, to the annoyance of the FCO,

---

13 NARA, ADD, RG 59: Bucharest to SecState, tel. 5785, 02.07.73; U.S. Mission OECD to SecState, tel. 7613, 26.03.75; U.S. Mission OECD to SecState, tel. 8777, 08.04.75; Gespräch des Bundesministers Genscher mit Präsident Ceauşescu, 05.12.75, *Akten zu Auswärtigen Politik der Bundesrepublik Deutschland, 1975*, Band II doc. 369, Munich, Oldenbourg, 2006, p. 1742-1754. For the general background see Mihail Ionescu, "Romania, Ostpolitik and the CSCE, 1967-1975", in Oliver Bange and Gottfried Niedhart (eds.), *Helsinki 1975 and the Transformation of Europe*, New York, Berghahn, 2008, p. 129-143 and Werner Lippert, *The Economic Diplomacy of Ostpolitik: Origins of NATO's Energy Dilemma*, New York, Berghahn, 2011, p. 62-100, 148-149.

14 NARA, ADD, RG 59: Bucharest to SecState, tel. 2395, 02.06.75. Gespräch des Bundesministers Genscher mit dem rumänischen Außenminister Macovescu, 08.07.76, *Akten zu Auswärtigen Politik der Bundesrepublik Deutschland, 1976*, Band I doc. 222, Munich, Oldenbourg, 2007, p. 1035-1038.

15 UKNA, PREM 16/638: record of a meeting between the PM and emissaries from President Ceauşescu on 23.04.75.

who deplored the dissemination of such messages via non-diplomatic channels.[16] More probably, Wilson's solicitude, also manifested through a certain relaxation of protocol,[17] was due to the wish to give the British economy as many chances as possible to exit from an awkward situation, by full exploiting the possibilities offered by the process of Détente. Indeed, in the course of a brief meeting at Chequers with Ceauşescu, who was on his way back from Latin America, on June 12, 1975, it was Wilson that urged for closer cooperation through Anglo-Romanian joint ventures of the kind operating in developing countries. Wilson also proposed sending Peter Shore, Secretary of State for Trade, to Bucharest at the earliest opportunity, despite recommendations to the contrary from the Department of Trade (DOT), who regarded the prospects offered by the Romanians as significantly less promising than they had been in the optimistic climate of a few years before.[18]

Although many people looked askance at Romania and even if the United Kingdom, like other Western countries, was constantly bombarded with requests for export loans on conditions that were tantamount to aid for development, in September 1975, not only did a British Prime Minister visit an East-European country other than the USSR for the first time since the Second World War, but Margaret Thatcher, the new leader of the Conservative opposition, also went to Bucharest. Mrs Thatcher jokingly told Ceauşescu that she had beaten Wilson to it in visiting him and her conversations with him and with his wife Elena (whom Mrs Thatcher found 'warm, relaxed, and agreeable')[19] lasted much longer than planned, leading Ambassador Petersen to remark that this was more than a mere courtesy visit.[20]

Ceauşescu affirmed that Anglo-Romanian cooperation had taken a leap forward under the Conservative government with the Viper licence and that he hoped things would go on equally well when Mrs Thatcher became Prime Minister. Naturally, these compliments did not spare the Romanian President Mrs Thatcher's cutting remarks when he pointed out that the military blockades should be eliminated and that true Détente was not possible under the shadow of terror:

---

16 UKNA, FCO 28/2549: Ashe to Bullard, 21.01.74.

17 UKNA, FCO 28/2750: Wright to Dales, 30.05.75.

18 UKNA, PREM 16/637: steering brief: origin and purpose of meeting, n.d.; record of a meeting between the PM and the President of Romania on 12.06.75; UKNA, FCO 28/2750: minute by Terry, 13.06.75.

19 UKNA, FCO 28/2751: Petersen to Cartledge, 04.09.75.

20 UKNA, FCO 28/2751: Petersen to Killick, 05.09.75.

> Comrade Nicolae Ceausescu: Would you trust a travelling companion who was armed?
> Margaret Thatcher: It depends. If I had a gun in my pocket and I knew how to draw it faster, I would. But it would be better if neither were armed.[21]

A couple of weeks later, Wilson also arrived in Bucharest and during his two-day visit he persistently returned to the question of joint ventures. A number of understandings were signed, one of them a ten-year industrial and technological cooperation agreement whereby the contractors undertook to create the most favourable possible credit conditions as an essential prerequisite for industrial collaboration. In a joint statement, the two parties also set a target to increase bilateral exchanges 2.5-fold by 1980.[22] Joint enterprises, seen as the best way for cooperation, along with what was happening in the aviation industry, would be emphasised *a posteriori* in order to foster Ceauşescu's personality cult, precisely when the regime was going through the most autarchic and oppressive phase while struggling to extinguish foreign debt.[23]

It was in this context, during a private meeting, that Ceauşescu broached the subject of the 614 to Wilson, explaining that Romania was in process of negotiating a joint venture with VFW-Fokker for the production of the 614 and a successful outcome was on the cards. Romania would, therefore, be interested in developing further cooperation with Rolls-Royce to produce the engine for the 614 under licence – this when only a few months before the Germans had assured Washington in a CoCom meeting that the M45H technology would not be transferred to Romania. Wilson pointed out that although such a project would raise serious problems in terms of financing and countertrade, his government and Rolls-Royce would be willing to consider a joint venture of this type. He made no mention of possible security problems since at that time the attention in this regard was focused on the afterburner of the Viper.[24]

---

21 ANR, CC, Foreign Relations, 1975_153: stenogramă primirii de către tov. N. Ceauşescu a liderului Partidului Conservator din Marea Britanie, 02.09.75, f. 6.

22 PREM 16/635: long-term agreement on economic collaboration and industrial and technological cooperation between the United Kingdom and Romania, 18.09.75; PREM 16/636: Bucharest to FCO, desp. 13, 18.09.75; note of a meeting between the PM and the President of Romania on 16.09.75 at 7:05 p.m.; UKNA, FCO 28/3426: visit of the PM to Romania, 16-18.09.75, n.d.

23 *O politică de pace, colaborare şi prietenie cu toate popoarele*, Vol. 1, Bucharest, Ed. Politică, 1983, p. 603-606.

24 PREM 16/636: note of a meeting between the PM and the President of Romania on 16.09.75 at 5:00 p.m.; NARA, ADD, RG 59: U.S. Mission OECD to SecState, tel. 8777.

This conversation should be looked at in the light of the difficulties that the 614 negotiations were encountering. Only ten days earlier, Ceauşescu had expressed irritation to the President of VFW-Fokker, Gerrit Klapwijk, saying that after years of negotiations, it was time to settle the issue of cooperation. Klapwijk, who seemed to be unaware of the 1973 Memorandum of Understanding between the FRG and Romania, affirmed that he had already been informed by Prince Bernhard of the Netherlands, honorary president of the board of the company, of Romania's dissatisfaction, but he explained that the deciding factor for the success of the project was that the German – rather than the Dutch – government should continue to finance it. Unlike Klapwijk, Ceauşescu seemed confident that Bonn would invest in Romania, especially in view of the country's huge trade surplus and, indeed, in March 1977, the Germans designated 50 million marks towards the joint company's marketing plan. There were, however, signs of discontent, also from Chancellor Helmut Schmidt, which substantiated Klapwijk's concerns. In particular, it was pointed out that Romania's partner was VFW-Fokker, so the German government was in no way responsible for the final outcome of the project.[25]

So, after breaking the ice with Wilson, from the autumn of 1975, the Romanians began to test the waters in the UK. To put it plainly, after the experience with the Viper and with a contract with China in the pipeline, Rolls-Royce's top management were unenthusiastic about the prospect of a new licence agreement with Romania. They would have preferred to sell the engines directly, perhaps with the Romanians producing some of the components, also for financial reasons. At this point, however, Bucharest was ready to activate a project to compete with the 614, most likely to use the usual trick of pitting the Westerners against each other and then seeing who would come up with the best offer. Be that as it may, HSA had no intention of resurrecting the idea of a licence to construct the 748. On the contrary, BAC was very keen to keep the 1-11 alive as long as possible, since the company's future in civil aviation depended on it; so the company offered the Romanians the possibility to build the entire Series 475 in their country.[26]

---

25 ANR, CC, Foreign Relations, 1975_158: stenogramă discuțiilor avute cu ocasia primirii de către tov. N. Ceauşescu a preşedintelui concernului aeronautic VWF-Fokker, 06.09.75, ff. 2-5. Deutsch-rumänisches Regierungsgespräch, 23.05.77, *Akten zu Auswärtigen Politik der Bundesrepublik Deutschland, 1977*, Band I, doc. 126, Munchen, Oldenbourg, 2008, p. 644-648; Aufzeichnung des Staatssekretärs Hermes, 09.12.77, *Akten zu Auswärtigen Politik der Bundesrepublik Deutschland, 1977*, Band II, doc. 360, Munchen, Oldenbourg, 2008, p. 1737-1739.

26 UKNA, BT 359/9: Dorman to Roberts, 23.06.76; UKNA, FV 14/86: Page to Kaufman, 21.07.76; Gerhard to Dorman, 22.11.76.

In May 1976, Ferguson Smith informed the Department of Industry (DOI) that BAC had devised a project for the creation of a production line in Bucharest parallel to the one in Britain. Over ten years, starting from 1978, the two lines would produce fifty planes, with the Romanian line becoming increasingly independent and sharing in the management of exports to third parties. Although the project looked fairly profitable for BAC, Whitehall officials were unenthusiastic. Firstly, why keep talking about fifty planes, as was the case for the 614, when the 1-11 was much bigger? Secondly, why should BAC get involved in a cooperation agreement for production and marketing, rather than simply sell the manufacturing licence? Thirdly, seeing that it was no secret that the difficulties with the VFW-Fokker joint venture were financial, the Romanian demands could have been burdensome or impossible to satisfy, as for example by obliging British Airways to purchase a certain number of 1-11s to ensure the profitability of the venture.[27] In short, if in the recent past internal considerations had prompted the British government to make a financial commitment in order to facilitate the sale of the five 1-11s to Romania, now, this very precedent – along with the particular circumstances of the upcoming nationalisation of the aerospace sector – indicated that it should do the opposite. As Anthony Warrington, head of the DOI Air Department, remarked:

> [BAC's] shareholders have withdrawn effective support and control from the civil side, so it is all being thrown at the Government. The managers' jobs are at stake as much as the workers'. The last Romanian deal the particular members of BAC's management made has been expensive, and much of it could fall on HMG through the underwriting contract. Thus our line is to welcome what has been achieved thus far [but] we cannot give any impression that the conditions laid down by the Romanians [...] should involve us in financial commitments.[28]

Whitehall's worries were justified. Romania's plans to enlarge and modernise the TAROM fleet were gainsaid by the drop in tourist demand, which was not even enough to utilise the aircraft already in service. To make sure of a paying load, the commander of civil aviation, General Aurel Răican, requested and obtained the authorisation to organise charter flights from Africa and Asia to Saudi Arabia for Muslim pilgrims.[29]

27 Ministerul Afacerilor Externe din România (MAER), Bucharest, 1976_528: Pațan la Popa, tel. No. 03/07658, 11.05.76; UKNA, FV 14/86: minute by Warrington, 01.06.76; minute by Jardine, 16.07.76.

28 UKNA, FV 14/86: minute by Warrington, 22.07.76.

29 ANR, CC, Administrative-Political Section, 1976_2: notă No. 21.580, 20.07.76, f. 133; notă No. H.3671, 26.07.76, f. 131.

The situation, which was at least partially due to the global economic situation, was exacerbated by the poor quality of tourism services offered in Romania, from vexatious measures towards visitors – such as the compulsory minimum daily exchange rate – and the lack of investment in the sector.[30] Also the commercial marketability of the 1-11 was dubious, as already demonstrated by the Romanians' insistence on a purchasing commitment from BA. The focus was on the Chinese, who, having bought the licence for the Spey 202, were likely to be interested in buying planes using the civil version of the engine from a friendly country. However, concrete results were contradictory and, when it came to the point, rather unfavourable.[31]

For its part, BAC clearly recognised the importance of Romania's on-going efforts to secure sales in different countries, especially in Asia, because this could help to keep the market open, thus allowing the possibility of further development of the 1-11. So, BAC opted for cooperation, in line with what had so far been achieved with the Romanians, firstly with a view to competing against the 614 and, secondly, in order to avoid being dragged into a proper joint venture. The last option would have meant investing fresh capital and a further deterrent was the experience of the French company Citroën.[32] The Letter of Intent, dated July 19, 1976, from the General Director of Tehnoimport, Tudorel Harabagiu, for the batch of fifty Series 475 aircraft, was an important result, in the sense that it officially recognised BAC as a competitor of VFW-Fokker. However, it left an enormous number of problems unsolved.[33]

In the first place, the Romanians insisted that fifty aircraft (the first five supplied by BAC and the last five produced entirely in Romania) were not enough for the project to be profitable. They spoke in terms of over 170 aircraft, a demand that BAC considered to be absolutely unrealistic. Equally absurd, in BAC's view, was Bucharest's underestimation of the time needed to complete the transfer of know-how. The Romanians, anyway, were not willing to give up the idea of forming a joint production company and they launched the usual offensive aimed at obtaining a strong reduction in the cost of the manufacturing licence and the royalties. Another problem was that one of the conditions stipulated by Bucharest was the possibility to produce the civil Spey engine, in line with

---

30 Tsantis – Pepper, *Romania*, *op. cit.*, p. 358-370.

31 UKNA, BT 359/9: Dorman to Roberts, 23.06.76; UKNA, FV 14/86: minute by Weaire, 21.01.77; UKNA, FV 14/87: minute by Weaire, 08.02.77.

32 On OLTCIT see Wiener – Slater, *East-West Technology Transfer*, *op. cit.*, p. 308-309.

33 UKNA, FV 14/86: Harabagiu to BAC, 19.07.76.

Ceauşescu's request about the M45H during Wilson's visit. Although in one sense Rolls-Royce had a preference for the Spey, which would have ensured much higher profit margins, on the other hand they thought that the Romanians would not be able to manufacture the engine under licence without massive technical assistance, which they had no intention of offering, lest it should deflect too many resources from other projects. They also thought that Bucharest's real target was the military engine. For Rolls-Royce, all these aspects could have compromised the company's relations with the Americans, which they considered to be far more important strategically. Rolls-Royce, therefore, was not prepared to draw up a licensing agreement for the civil Spey. This provoked criticism from BAC, who saw it as a hindrance to their negotiations.[34]

At the end of 1976 things were still rather nebulous and Whitehall officials were visibly sceptical about the prospects. They believed that the British government should keep out of the negotiations, reiterating that no special concessions would be made (for sure British Airways would not be forced to purchase the 1-11) and always insisting that any agreement would have to be commercially viable. All they were willing to do for BAC was review the 1975 underwriting agreement, so that half of the profits of a possible continuation of the production of the 1-11 would go to BAC – instead of being used to pay off the debt contracted with the government for the reopening of the line, as had happened in the case of the original batch of the five aircraft for TAROM – thus extending the advantages of the new Anglo-Romanian project under negotiation.[35]

The Romanians soon forced BAC to accept the principle of the joint venture, albeit mitigated by a legal clause that limited financial liability to ten million pounds – namely, the amount BAC was planning to invest in the project, only half of it in cash – to be deducted from the twenty million that Romania had been asked to pay as a licence fee. The Deputy Minister for Foreign Trade, Stanciu, exerted considerable pressure in this regard, even complaining about Rolls-Royce's obvious lack of enthusiasm, when cooperation on both versions of the Spey, civil and military, would have

34 MAER, 1976_528: Popa la Mănescu, Oprea, Paţan şi Avram, tel. No. 050363, 09.07.76; UKNA, BT 359/9: minute by Dorman, 15.07.76; Dorman to Jardine, 16.07.76; UKNA, BT 359/21: Gerhard to Dorman, 26.08.76; Dorman to Gerhard, 27.10.76; Gerhard to Dorman, 04.11.76; UKNA, FV 14/86: note for the record – BAC 1-11 for Romania, 15.09.76; note of a meeting at DOI on 15.11.76; note of a meeting with BAC on 14.12.76; minute by Elias, 11.01.77.

35 UKNA, FV 14/86: Lightman to Petersen, 07.10.76; minute by Octon, 01.11.76; Jardine to Ferguson Smith.

greatly strengthened BAC's negotiating position against VFW-Fokker, which was still the preferred option.[36]

The ROMBAC project envisaged the production of one hundred planes over a ten-year period: the first five would be purchased by BAC, but with the components manufactured by the Romanians in accordance with the 1975 cooperation agreement. The transfer of know-how would apply to the Series 475 (possibly extending to the Series 500), and it would last five and a half years. The twenty-first plane would be 60% manufactured by Romania and from the twenty-fifth aircraft, assembly would take place in Romania, according to the principle of parallel production lines. BAC was to retain design authority – so the Romanians would not be able to modify the airplane autonomously – and some components, such as the undercarriage, would continue to be produced in Britain.[37]

In addition, BAC agreed to ensure a five-year credit on 85% of the value of what was purchased in the United Kingdom. Such conditions were normal for this type of export credit (no longer the 90% agreed under the 1975 transaction), but less favourable than those offered by the Germans (with an eight-and-a-half-year deadline), but the hostility of the ECGD and the DOT was palpable. Romania's economic prospects were not considered alarming, partly because the debt servicing liabilities decreased even more rapidly after 1980, but Romania was one of the ten countries where the United Kingdom was financially most exposed. It was unfeasible to finance the ROMBAC project, the Spey and possibly the M45H simultaneously, simply because there would be no money left over for other industrial initiatives that were considered, rightly or wrongly, more dependable than the aviation projects. Therefore, it was necessary to decide which to finance: the ROMBAC or the Rolls-Royce engine licences. In any case, if the ROMBAC project was chosen it would be for no more than fifty aircraft and payment would be in dollars, in order not to interfere with the ECGD commitment to reduce sterling borrowing after the serious currency crisis that the United Kingdom had just gone through.[38]

---

36 UKNA, FV 14/87: new elements in revised proposals for BAC 1-11 collaboration, 30.01.77; note of a meeting held at DOI between Mr. Kaufman and Mr. Stanciu on [24].01.77; Ferguson Smith to Harabagiu, 03.02.77; minute by Warrington.

37 UKNA, BT 359/9: ROMBAC 20C.1, 11.02.77.

38 UKNA, FV 14/86: Skipworth to ECGD, 16.11.76; UKNA, FV 14/87: Dorman to Terry, 04.03.77; UKNA, FV 14/88: EG (77) 2nd meeting of the Export Guarantee Committee, 03.03.77.

This state of affairs had at least one advantage. Indeed, both Ambassador Secondé and BAC – either directly or through Pop – told the Romanians that there were not enough funds to finance both the M45H licence for the VFW-614 and the ROMBAC project, suggesting that it was in Bucharest's interest to decide in favour of the latter, putting to good effect the background of experience accumulated in the bilateral cooperation. Furthermore, the negotiations on the Rolls-Royce engines, started only in November 1976, and followed in a perfunctory manner, clearly demonstrated that Rolls-Royce had no intention of committing itself to two simultaneous licence contracts and that its preference was for the Spey. Privately, Rolls-Royce was hoping for the definitive closure of the 614 in order to free itself from the expensive M45H programme. However, although the Romanians recognised the impossibility of proceeding with both projects simultaneously, they seemed to be set on going ahead with the policy of running two projects in parallel, but with different time schedules, especially after Bonn had made a settlement of 50 million marks in March 1977. They assured both the British and the Germans that their project would take off first.[39]

After being very cordially received by Ceauşescu in person, the managers of BAC – which by then had been taken over by British Aerospace (BAe) – signed a protocol that modified the programme even further, on May 28, 1977. They were now talking about eighty-two planes, with Romania assembling the first complete plane in 1980 and, by 1985, the whole airframe autonomously. Electrical and electronic systems were to be bought or produced in collaboration on conditions to be established at a later date. The Romanians also demanded that at least half the value of the British supplies be covered by countertrade. Lorenz Schomerus, Secretary of State to the German Ministry of Economic Affairs, observed that a lot of documents had also been signed with VFW-Fokker. Schomerus told the British Ambassador to Bonn that this would have been surprising anywhere else in the world, but not in the Balkans. On July 2, 1977, the Romanian-German company was launched. It was to supervise the construction of one hundred VFW-614s, half of which were to be exported to countries as different as Communist China and General Pinochet's Chile. Pop's contacts in Tehnoimport further confirmed that both projects would go through, but not at the same time, and that it had not been

39 MAER, 1976_528: Popa la MAER, tel. No. 050824, 01.11.76; UKNA, FV 14/88: Dorman to Terry, 25.03.77; minute by Rogers, 28.03.77; Dorman to Gerhard, 30.03.77; Secondé to DOI, creda 42, 25.04.77; UKNA, FV 14/89: Gerhard to Edmonds-Brown, 05.07.77.

decided which would start first, though Pop seemed optimistic for the Spey, given Romania's interest in the military version.[40]

Although the protocol was simply a declaration of intent, it was nevertheless an important step forward, insofar as it at least established the number of planes to be built under the project. With this in view, a framework agreement was signed on July 20, 1977, to the satisfaction of BAe. In the first place, the idea of the joint production company was abandoned in favour of a less formula, whose activity would consist only in identifying possible market outlets but without being involved in the negotiation of contracts. In the second place, the time schedules for the industrial plan were revised, making the length of the period necessary for the transfer of technological know-how more realistic and extending the time for the supply of components from the United Kingdom from five years to seven. In exchange, BAe threw into the cooperation mix the freight carrier version of the 475 and the possibility of adding the rear ventral door for loading/unloading. This would satisfy the requirements for use by the Romanian armed forces. However, many thorny issues remained unresolved: first, the number of aircraft that Romania would buy from BAe, who were offering five, while the Romanians did not want to buy more than two; secondly, the questions connected with the negotiations with Rolls-Royce and the 614 project with VFW-Fokker still had to be sorted out.[41]

In fact, negotiations on the engines moved slowly, with the Romanians persisting in wanting both the Spey and the M45H. While the delay may have worried BAe, it suited Rolls-Royce. As Pepper, Vice President of the company, pointed out to the DOI, it was better to aim at selling the Speys rather than just the licence as this would increase profit margin. Once the airframe projects had been realised, the more time passed, the less feasible the transfer of knowledge would become, making it necessary to buy the engines needed to complete the planes from the United Kingdom. What is more, this would have suited the British government not only indirectly in economic terms, but also directly for political reasons, because selling the complete Spey would be no problem in the CoCom area.[42]

40 UKNA, FV 14/88: Wright to DTI, creda 100, 01.06.77; protocol signed on 28.05.77; Dorman to Gerhard, 01.06.77; Secondé to DOT, creda 58, 04.07.77; UKNA, FV 14/89: Bowen to Rawlinson, 07.07.77.

41 UKNA, FV 14/89: framework agreement between BAC and GAB signed on 20.07.77; Brown to Gerhard, 20.07.77; note of a meeting held at DOI on 04.08.77; UKNA, FV 14/90: Ranson to Fordham, 21.09.77.

42 UKNA, FV 14/89: minute by Chilton, 28.07.77; Brown to Gerhard, 03.08.77; minute by Gerhard, 05.08.77; minute by Gerhard, 16.08.77; Gerhard to Brown, 16.08.77.

The Romanians seemed determined to proceed with both projects and, towards the middle of August 1977, they ordered fifty Speys and fifty M45Hs to be delivered between 1979 and 1981. Since the Treasury and the ECGD had agreed not to provide financial cover for the supply of ROMBAC components after 1983, given the uncertainty regarding numbers and especially prices – even though the DOI pointed out that in this way two years of the project would remain uncovered because of the new time schedules set by the framework agreement – the accumulation of credit for the airframe component and the two engines could now be accommodated under the general agreement with Romania. In spite of this, partly to limit the risk, partly to favour ROMBAC in line with the preference expressed by the DOI, the ECGD rejected the project submitted by Lazard for a credit line intended to cover the British elements of the 614 project. On the other hand, it offered the Romanians a loan that would cover only 80% of the fifty M45Hs (compared with the standard 85% conceded for the ROMBAC).[43]

Furthermore, to the President of the Romanian Bank for Foreign Trade, Vasile Voloşeniuc, who, on the day when the ROMBAC framework agreement was signed, had received instructions to contact German and British banks and financial institutions in order to obtain the necessary financing to launch the 614 project, the ECGD stated explicitly that it was very unlikely to intervene in either of the projects and it expressed concern about the possible negative effects of the 614 on the ROMBAC's fate. In his report to Paţan, Voloşeniuc specified that for the joint company with VFW-Fokker to take off, it would be necessary to obtain financial support from Bonn:

> In the course of unofficial discussions, the directors of certain banks have pointed out that the West German authorities, though not particularly interested in supporting exports, for what concerns the aviation industry, find themselves in a different situation, because without the support of the government, the industry would be in serious economic difficulties and would create serious political and social problems. In this climate, it has been suggested that the Romanian authorities should insist on the best financial terms from the West German authorities, with a greater use of subsidies that the government makes available to the West German aviation industry.[44]

43 UKNA, FV 14/89: Ranson to Forsham, 22.07.77; Lowton to Kelley, 08.08.77; UKNA, FV 14/90: minute by Chilton, 19.08.77; note of meetings held in Mr. Cotterill's room on 22.08.77.

44 ANR, CC, Economic Section, 1977_28: raport No. S/14068, 25.08.77, f. 19.

However, this was exactly what the Bonn government did not want. In autumn 1977, while BAe went ahead with the ROMBAC negotiations and the Romanians were coping with BN's second collapse,[45] VFW-Fokker was itself on the verge of bankruptcy. A recovery plan was set up to the tune of half a billion marks to limit the social repercussions of the company's crisis and it was proposed that VFW-Fokker should merge with MBB (Messerschmitt-Bölkow-Blohm), but the first victim of the restructuring process was the 614. The Germans informed Bucharest that the production line would be closed down by the end of the year in the absence of realistic trade outlets in the West. At this point, the Romanians launched a desperate attempt to save the project by finalising the purchase of three 614s by TAROM and pursuing all possible options to keep the company going, especially in the light of a possible order of ten planes from the American company FedEx, and to take advantage of the planned visit of the Chancellor to Bucharest to persuade him to untie the purse strings.[46]

Schmidt, who was already less than enthusiastic about the visit, refused to budge. After a long series of gripes about how the world economy had changed and how the industrialised countries would no longer be able to extend heavy loans to the socialist countries on favourable conditions, and that private banks should take on the moral risk of investments without asking for government help, the German Chancellor condemned the Dutch management of the VFW-Fokker question. Ceauşescu insisted that the company should be allowed to continue with its project, on the grounds of FedEx's interest and sales possibilities thanks to the good relations that Romania enjoyed with many countries in the world, but Schmidt retorted that the American contract was "unreliable" ("*ein windiger Vertrag*"), because it contained an escape clause and, in any case, FedEx would have only rented the aircraft, not bought them. As for the rest, he wished Ceauşescu good luck.[47]

---

45 UKNA, FV 14/91: notes of BAC delegation's visit to Bucharest on 06-08.11.77.

46 ANR, CC, Economic Section, 1977_28: notă cu privire la producerea avionului Fokker-614, 29.12.77, ff. 21-22; UKNA, FV 14/91: minute by Octon, 09.11.77; NARA, ADD, RG 59: Bonn to SecState, tel. 20866, 16.12.77. Aufzeichnung des Staatssekretärs Hermes, 09.12.77, *Akten zu Auswärtigen Politik der Bundesrepublik Deutschland, 1977*, Band II, doc. 360, *op. cit.*; Mark Lambert, "Germany's Aerospace Industry", *Flight International*, 28.04.78, p. 1277.

47 Gespräch des Bundeskanzlers Schmidt mit Präsident Ceauşescu, 06.01.78, *Akten zu Auswärtigen Politik der Bundesrepublik Deutschland 1978*, Band I, doc. 3-4, Munchen, Oldenbourg, 2009, p. 12-41.

With the 614 off the scene, the 1-11 remained – ten years after the purchase of the first aircraft – the only concrete opportunity for the Romanian aviation industry to get on the commercial jet ladder. In this new climate, BAe's insistence on speeding up the negotiations corresponded to Romania's willingness, in early January 1978, to send to London Boldur, the Vice Minister of the Machine-Building Industry, and Bucur, General Manager of GAB (Grupul Aeronautic Bucureşti), the industrial complex where the production line would be set up. It is significant that Pop, underlining the importance that this meeting could have, drew attention to the fact that any possible agreement, given the current situation, would have to make economic sense for Britain, and not be a vehicle for more or less secretive public subsidies, in the way that the 1975 transaction had been, or even more so, the recent Anglo-Polish agreement on shipbuilding.[48] Without doubt, the negotiations would clash with the preparations for Ceauşescu's state visit to the United Kingdom that was to take place in mid-June 1978.[49]

The failure of the 614 soon forced Klapwijk to resign amidst bitter controversy. It also brought about the final split between Fokker and VFW. Fokker continued to produce its F.27s and F.28s and it joined the F-16 fighter programme. VFW merged with MBB, working on the Airbus consortium and the construction of the Tornado.[50] It seems astonishing that a country like Romania, devastated by the earthquake of March 1977 and still producing only 35% of a small aircraft like the Islander, could seriously be thinking of embarking on two huge projects like the ROMBAC and the 614 more or less simultaneously. Throughout 1978, Romania continued to insist that the 614 should not be permanently shelved, but relocated to Romania. Thus, although back in April 1978, Vice President Pepper had pointed out that Rolls-Royce did not intend to negotiate on the M45H until Bucharest provided evidence that it had obtained rights over the plane from the Germans and that it was able to cover the costs of start-up and product support, for months the issue continued to be a serious hindrance to the Spey negotiations.[51]

---

48 On the Anglo-Polish shipbuilding deal see Leon Zurawicki, *Multinational Enterprises in the West and East*, Alphen aan den Rijn, Sijthoff & Nordhoff, 1979, p. 183-185.

49 UKNA, FV 14/91: Secondé to DOT, creda 98, 20.12.77; minute by Warrington, 21.12.77; minute by Warrington, 22.12.77.

50 "Klapwijk Leaves VFW-Fokker", *Flight International*, 27.05.78, p. 1587; "The VFW-Fokker Divorce", *Flight International*, 01.03.80, p. 613.

51 UKNA, FV 14/92: minute by Maynard, 11.04.78; UKNA, FV 14/93: minute by Maynard, 20.07.78.

On October 26, 1978, Pepper personally told Bucur that Rolls-Royce had finally decided to cancel the M45H and that he would send an official communication to this effect, but the GAB General Manager expressed doubts as to whether a simple letter from the British company would be enough to convince his government, and, above all Ceauşescu, that the matter was really closed. In fact, it was only the direct confirmation of Keith during one of his trips to Bucharest that finally convinced the Romanian president that the ROMBAC was the only alternative left.[52] It has been noted that Germany did not really want to cooperate with a Socialist country in a large aviation project, basically for security reasons regarding the possible theft of technology by Romania.[53] This problem was not underestimated even by the British, but that should not obscure the fact that, from both the economic and the political point of view, it was extremely hazardous and unappealing to embark on such a challenging project with a country that seemed to have lost all sense of proportion. The difference between the British and the Germans lay mainly in the fact that the management of the former BAC was right to consider the ROMBAC strategic for its future as an industry in the context of the new equilibrium resulting from the BAe takeover.

5.2 Queen Elizabeth II's Guests

The idea of a visit to the United Kingdom by Ceauşescu was first mooted in 1973, when Anglo-Romanian aeronautical relations took off again after their previous crisis. At the beginning of 1974, the British were told that Ceauşescu wished to be received by the Queen on a state visit, rather than by the British government on a simple routine visit, even though the former would take longer to organise.[54] Ceauşescu's preference for a high profile, ceremonial visit was in line with his narcissistic personality and the need for political propaganda both at home and abroad. Multiple direct contacts with the most powerful figures in the world were seen as important for the Conducător's personality cult. The fact that the first state visit to the United Kingdom by a Warsaw Pact leader was planned for mid-1978 raises two questions. First, how should an event that gave visibility and legitimacy to an increasingly repressive

---

52 UKNA, FV 14/158: note of a meeting between RR and Mr. Bucur on 26.10.78; Maynard to Mays, 03.11.78; visit of Sir Keith and Mr. Pepper to Romania, n.d.

53 Liviu Ţăranu, „Afacerea Fokker“, *Magazin Istoric*, No. 487 (2007), p. 49-52.

54 UKNA, FCO 28/2549: Bucharest to FCO, tel. 34, 07.02.74; UKNA, FCO 28/3080: minute by du Boulay, 13.07.77.

regime be managed? Secondly, what was the link between the visit and the ROMBAC negotiations?

Regarding the first point there was an element of embarrassment. For example, after a briefing with diplomatic correspondents during which they clearly expressed uneasiness at the idea of Ceaușescu appearing in close association with Queen Elizabeth II, the Foreign Secretary, David Owen, commented that the journalists were right to express regret about the plan.[55] Things were made worse by a number of diplomatic horror stories about the behaviour of Ceaușescu and his wife, such as the couple's rooms at the Belgian Royal Palace being searched by the Securitate when they were the guests of King Baudouin; the demand for a double bed to be fitted on the plane that was to fly them in just two hours from Caracas to Ciudad Guyana; a scene in New York because the Police Department had not banned a demonstration against Ceaușescu by Hungarian immigrants because of the treatment of the Magyar minority in Transylvania. Faced with such excesses, disorganisation and capricious pretentions, the American diplomats, veterans of one of these visits in April 1978, suggested that the key to success was to make Elena happy – because if she was happy, then so was her husband – and to heap praises on Ceaușescu's statesmanship in order to keep the wheels oiled during the visit.[56]

These situations, however, remained fraught with potentially embarrassing predicaments. For example, when Wilson visited Romania in 1975, in recognition of Elena's meteoric rise to the top of the RCP, the FCO suggested that she should receive an honorary degree – perhaps from Bradford University, of which Wilson was Chancellor – during the state visit, rather than be invited to visit the United Kingdom separately.[57] In early 1978, however, Wilson announced that Bradford was not willing to fall in with this plan. This sparked a sort of 'degree hunt' in which both the FCO and the Romanian diplomats played their parts. The Royal Society, the University of Southampton, Imperial College London and Imperial Chemical Industries all got involved. The whole scenario was based on the false supposition – upheld by British diplomats in Bucharest – that Elena

---

55 UKNA, FCO 28/3427: minute by Fergusson, June 1978; UKNA, FCO 28/3428: minute by Walden, 12.06.78.

56 UKNA, FCO 28/2549: Hanbury Tanison to Bullard, 08.11.72; Gibbs to Hunter, 19.09.73; UKNA, FCO 28/3425: Burns to Lambert, tel. 63, 20.04.78; Secondé to de Boulay, 24.04.78.

57 UKNA, FCO 28/2549: Fletcher to Roberts, 18.10.74; UKNA, FCO 28/3080: Sullivan to Shea, 05.12.75.

was actually a talented researcher in Organic Chemistry. This was despite the fact, borne out by witnesses, among them the physicist and leading authority on semi-conductors, Radu Grigorovici, that the Romanian First Lady was a dimwit and that awarding her academic degrees only served to give credence to the propaganda of the dictatorship.[58]

As the number of refusals mounted,[59] attention turned to the Royal Institute of Chemistry who were prepared to offer Elena a fellowship, along with a scroll of parchment, and the Polytechnic of Central London (PCL), which had links with the Polytechnic of Bucharest and the Ştefan Gheorghiu Academy – the training ground for party members – mainly through its School of Management Studies and which was keen to take advantage of the State visit to strengthen these contacts. The Vice Chancellor of the PCL had already written to Leonte Răutu, the Žhdanov of Romanian Stalinism[60] and Rector of the Ştefan Gheorghiu Academy, saying that he was willing to host the presidential couple only briefly. Later, as a solution to the problem, he offered Elena an honorary professorship. So, Romania's First Lady was able to spend a morning picking up her academic titles, snubbing ICI, the Royal Society and Imperial College (where her son Valentin had been a student for two years) even though they were ready to welcome her, leaving half the afternoon free for shopping in London.[61]

Though the British did not justify the depressing picture of nepotism and 'Eastern despotism'[62] offered by the Ceauşescu regime, they at least interpreted it with an ill-concealed sense of historic superiority as being of economic interest. For example, Ambassador Secondé, after describing Ceauşescu as the most ruthless dictator in Eastern Europe, shrouded in a sinister personality cult, concluded:

> But the standards against which these judgements are made are those of Western democratic society. The real background is that of a Balkan country, with no parliamentary tradition of any depth, but a long history of rule by

58 UKNA, FCO 28/3080: Burns to Figgis, 17.10.77; Prendergast to Falkender, 22.12.77; UKNA, FCO 28/3423: Burns to Lambert, 19.01.78; minute by Prendergast, 06.02.78; minute by Lambert, 13.02.78; Secondé to Scott, 06.03.78; UKNA, FCO 28/3428: resumé of a meeting with Prof. Radu Grigorovici on 26.05.78.

59 FCO 28/3424: Pitts to Secondé, 29.03.78; Scott to Secondé, 07.04.78.

60 Vladimir Tismăneanu and Cristian Vasile, *Perfectul acrobat. Leonte Răutu. Maştile răului*, Bucharest, Humanitas, 2008.

61 UKNA, FCO 28/3080: Burlin to Răutu, 06.12.77; UKNA, FCO 28/3424: Burlin to Wolstenholme, 07.04.78; UKNA, FCO 28/3425: Lambert to Norman, 20.04.78; UKNA, FCO 28/3426: Burlin to Lambert, 03.05.78; minute by Lambert, 09.05.78; minute by Stainton, 31.05.78.

62 UKNA, FCO 28/3080: Richards to de Boulay, 06.12.77.

> petty despots. […] Romanians have survived, not because of their military strength – they are too small – but because of their deviousness and tenacity […]. Ceauşescu fits neatly into this historical pattern. […] from the point of view of commercial and political prospects for Britain (and the West in general), I believe that we are lucky to have President Ceauşescu in the saddle. He is courageous, independent-minded and prepared to treat all fields with the West. On the whole, his polices have been beneficial to us so far.[63]

Trade had always conditioned bilateral relations with Romania. Now, the crisis of Détente in Europe, the fact that the Bucharest regime had lost its appeal and the state of the British economy meant that once again trade was the main focus of the state visit for the United Kingdom. Prime Minister Callaghan repeatedly urged Whitehall officials to exploit Ceauşescu's visit to the maximum in order to finalise the big contracts already being negotiated in the aviation sector and in the chemical and iron and steel industries. On government instructions, even the welcoming speech of the Lord Mayor of London was to centre on the commercial aspect of bilateral relations.[64]

In this context, and after the virtual bankruptcy of VFW-Fokker, for the British the state visit became the instrument for resisting pressure from Romania to secure more and more favourable conditions, while the Bucharest government tried to take advantage of the United Kingdom's obvious interest in the ROMBAC, with a view to obtaining concessions in other sectors, especially textiles and iron and steel. If BAe – though maintaining that it could rapidly finalise the licence agreement – actually intended to put off the negotiations on counter-trade till just before the state visit in order to put pressure on the Romanians, during the visit to the UK of the Foreign Minister, Macovescu, all the people he met insisted on the need to substantiate Ceauşescu's trip with the signing of important contracts. In particular, it was suggested that Ceauşescu might visit Bristol, where both BAe and Rolls-Royce had large factories. Naturally, regarding textiles and iron and steel, there was no room for manoeuvre, with the British government caught between EEC prerogatives and the need to protect jobs. As a result, Romanian requests for counter-trade would have to be reasonable, as Boldur was rather brusquely informed during his visit to London, which, contrary to Pop's expectations, was inconclusive.[65]

---

63 UKNA, FCO 28/3407: Secondé to Owen, diplomatic report No. 194/78, 24.04.78.

64 UKNA, FCO 28/3080: Prendergast to Cartledge, 20.10.77; Cartledge to Prendergast, 24.10.77; UKNA, FCO 28/3426: Scott to Ellis, 15.05.78.

65 UKNA, SUPP 29/297: note of a meeting with BAC on 25.11.77; UKNA, FCO 28/3077: note of the Secretary of State's meeting with the Romania Foreign Minister

In fact, the great dispute had begun over the price of the 1-11s that TAROM was to buy and over the manufacturing licence, with the positions of the two sides being very far apart. It should be pointed out that although the British were ready to negotiate, they took a firm stand. Of course, this did not mean that the ROMBAC had in some way become less important – quite the contrary. The sale of the eleven BAC 1-11s to Romania over ten years had guaranteed the survival of the aircraft and had made a significant profit for BAC. Moreover, it was obvious that the project would serve to keep production going in the United Kingdom, allowing these to survive up until the launch of a new generation of carriers. In any case, both the BAe negotiators and the Whitehall officials were adamant that the ROMBAC project should be economically advantageous for the United Kingdom. A new underwriting agreement was out of the question and the wish to conclude the negotiations in time for the state visit, thus favouring Bucharest's independence from Moscow, was subordinated even by the FCO to the need to define commercially favourable terms. Naturally, this attitude came as a surprise to the Romanians, who were not even in a position to resort to the tactics of playing off the Westerners against each other, with the Germans out of the game.[66]

There were two problems connected to the airframe negotiations. The first problem was the certification for the ROMBAC and for the engines. Without the CAA certification for aircraft manufactured in Romania, the market prospects would be drastically reduced, but from the series of meetings starting at the end of 1976, it had emerged that BAe did not trust in the project sufficiently to previously guarantee the same 'virtual equivalence' for planes manufactured in Romania and those produced in the United Kingdom. The Romanians seemed to have understood the gravity of the problem and it looked as if they wanted to establish an authority of certification pursuing the British Civil Airworthiness Requirements, however the issue remained unsettled. Furthermore, the British Airworthiness Requirements Board noted that in a country like Romania there could be no certification authority that was truly independent of the government so they asked BAe and Rolls-Royce to guarantee that the Romanian products

on 14.11.77; Cartledge to Prendergast, 16.11.77; record of conversation between the F&C Secretary and the Romania Foreign Minister on 15.11.77; UKNA, FV 14/91: note of a meeting held by Mr. Kaufman with the Romanian delegation to discuss the BAC 1-11 collaboration on 06.01.78.

66 UKNA, BT 359/10: Gerhard to Hill, 27.07.77; UKNA, FV 14/91: minute by Warrington, 20.01.78; Sutherland to Warrington, 24.01.78; Warrington to Ferguson Smith; Bucharest to DOT, creda 22, 03.02.78; situation report for the Rt. Hon. Dell, 14.02.78.

would respect the manufacturers' indications. The problem, then, stemmed from the unusual fact of having to deal with a manufacturer that did not hold the Type Design certificate, which remained with BAe, the only design authority overseeing the effects of the transfer of technological know-how. So, leaving aside Romania's wish to acquire its own competence regarding certification, for the time being the understanding was that the CAA would act on the basis of the guarantees provided by BAe.[67]

The second problem was the delay in the negotiations with Rolls-Royce for the engines compared with those with BAe for the airframe. This was partly because the Romanians continued to insist doggedly on bringing up the question of the M45H. During the talks, Bucharest had scaled down their ambitions regarding the Spey. This was due to the lack of skilled labour and, once again, to the absence of the necessary certification. Negotiations, therefore, centred on the extension of Turbomecanica's activities as a subcontractor of Rolls-Royce, with a plan to produce 250 engines (fifty of which would be supplied directly from the United Kingdom) over fifteen years. Initially, the Romanians would only be in charge of maintenance of the TAROM fleet's Spey engines. Later they would take increasing responsibility for manufacture, eventually covering two thirds of the British content (roughly half the engine production). For reasons of security and certification, there were some manufacturing operations that the Romanians could not take on.[68]

The technology required for producing engines, especially for civil purposes, was certainly the most delicate aspect from the point of view of CoCom checks. In this case the negotiations were further complicated by poor communications between the various British operators involved. Whereas the DOI supported Rolls-Royce, the DOT was convinced that transfer of all technological know-how to the Romanians was out of the question. The effects of the misunderstanding were quite amusing when Edmund Dell, the Secretary of State for Trade, went to Bucharest in 1978 for the purpose – on the Prime Minister's instructions – of underlining the importance of substantiating the state visit with the signing of some important trade agreements:[69]

67 UKNA, DR 8/27: Brady to Standley, 03.11.76; Brady to Standley, 07.02.77; Brady to Standley, 18.10.77; Morrison to Standley, 03.01.78; note of meetings at the CAA Airworthiness Department on 2-3.02.78; Morrison to Guilbert, 25.04.78; minute by Guilbert, 19.05.78.

68 UKNA, BT 359/21: minute by Octon, 21.03.78; UKNA, FV 14/92: minute by Waire, 07.03.78; minute by Haynes, 20.04.78.

69 UKNA, BT 359/21: Terry to Maynard, 07.03.78; UKNA, FCO 28/3442: Baker to Cartledge, 10.03.78.

> [...] at Paţan's first meeting [Dell] referred to the 'insuperable difficulties' which the manufacture of Rolls-Royce engines in this country would cause us. This stopped the interpreter in his tracks and the statement was repeated. At the same time, Don Pepper (Vice Chairman of Rolls-Royce) kicked me sharply on my right ankle and said that he did not think he could accept this. I mentioned my disquiet to the Minister in the car afterwards and he simply said that he had read straight off the brief. [...] John Rogers (the main day-to-day negotiator for BAC) had worked up a considerable head of steam: he approached me and said that they were now gravely embarrassed and what was I going to do about it? Meanwhile Don Pepper sneaked off to a telephone and rang up Warrington (Department of Industry) to ask 'in Irish' what the position was. Apparently Warrington assured Pepper that he was right.[70]

Dell was forced to resort to a cover-up strategy to close this unfortunate incident, but Pungan – who had become a presidential advisor with the rank of minister and a key person in the aviation negotiations – recalled that the breakdown of negotiations on the 748 should serve as a cautionary tale for everyone. More explicitly, the following month, he told the Secretary of State for Industry, Gerald Kaufman, that the ROMBAC negotiations would not be concluded until an agreement had been reached with Rolls-Royce.[71]

As if all this were not enough, the Romanians created further problems. It should be remembered that financial cover from the ECGD was in dollars. This was mainly because, with super-inflation and the currency crisis of 1974-76, using the dollar helped to make export prices more competitive and to stabilise the pound. Since that time, however, the weakness of the dollar had started to cause difficulties in negotiations that had been dragging on for about two years. BAC's prices were calculated in pounds; initially, they had been proposed to the Romanians at an exchange rate of $1.60 dollars to the pound in 1976. In mid-1977, the exchange rate had reached $1.74 and by the end of the year it was $1.91. It is true that the Romanians had already agreed to an adjustment, however they were not willing to do so a second time and they asked for the financial cover to be in pounds sterling. The BAe negotiators maintained that the move to sterling would not solve Romania's problems because the leu had a fixed exchange rate with the dollar that had been administratively set. On the other hand, the ECGD and the Treasury, though under pressure, were very reluctant to accept financing in sterling, not because of the cost

70 UKNA, FCO 28/3442: Secondé to Scott, 08.03.78.

71 UKNA, FCO 28/3442: note of a meeting with Mr. Pungan on 02.03.78; UKNA, FV 14/93: note of a meeting between Mr. Kaufman and Mr. Pungan on 26.04.78.

of the operation itself, but because they suspected that BAe was using the Romanians as a shield in order to attack the currency finance policy about which the company already had reservations. Now, believing that Romania had no choice other than the ROMBAC, the ECGD and the Treasury were determined not to make any concessions.[72]

Faced with this opposition, BAe wrote to Secretary of State for Industry, Eric Varley, complaining that the ECGD's decision would mean more delays, because the Romanians would refuse to be exposed to a long-term exchange rate risk, thus jeopardising the signing of the agreement on the occasion of the forthcoming state visit. Varley wrote to Dell asking for more flexibility on the issue, not only for the value of the contracts themselves, but especially for the impact on employment in the United Kingdom. Dell sent a letter to the Chief Secretary to the Treasury, Joel Barnett, supporting the DOI's requests with the argument that, if the contracts were not closed in time for the state visit, then the Romanians would no longer have the incentive to restrict their demands for concessions. It was only at this point, less than a week before Ceauşescu's arrival, that the Treasury back-pedalled by agreeing to one third of the planned capital for the whole programme in sterling.[73] In his reply to Dell, Barnett explained that:

> [...] knowing how credit-hungry Eastern European countries are, it is vital that we retain control over the volume and nature of the financing to be provided for/ in respect of a programme of activity which will take a number of years to complete. We still need to finance as much as possible of our export credit on a foreign currency basis [...]. We must therefore make strenuous efforts to get the Romanians to accept that the balance of the finance needed for the BAC 1-11 project shall be provided in dollars.[74]

The other problem that arose had already been hinted at by Ceauşescu's insistence, during Dell's trip to Bucharest, that the ROMBAC should not be a mere business transaction, but should be part of a broader picture of cultural and technological collaboration.[75] Thus Pungan, who arrived in London in April 1978 to complete the preparations for the state visit, presented Dell with a draft intergovernmental agreement on aerospace

---

72 UKNA, FV 14/91: minute by Chilton, 15.02.78; UKNA, FV 14/92: Beach to Kelley, 23.02.78; minute by Chilton, 24.02.78; Hanby to Cotterill, 07.03.78; Kelley to Cotterill, 20.03.78; Crâiniceanu to Cotterill, 03.04.78.

73 UKNA, PREM 16/1838: Varley to Dell, 19.05.78; Dell to Barnett, 05.06.78; UKNA, FV 14/93: Page to Varley, 12.05.78; Breach to Kelley, 09.06.78.

74 UKNA, PREM 16/1838: Barnett to Dell, 09.06.78.

75 UKNA, FCO 28/3442: summary record of conversation between the President of Romania and the Secretary of State for Trade on 02.03.78.

cooperation. The head of the DOT Commercial Relations and Export Department, Craig Octon, judged this to be completely unacceptable insofar as it did not take into account British commitments to CoCom; it involved the Government in what were supposed to be commercial transactions with businesses as the key players and it demanded, in essence, that Romanian products should be guaranteed an international certification. By this time the DOT had adopted a decidedly negative attitude and it was only in the light of the obvious difficulties in the negotiations that Secondé's recommendation was followed. He wanted the United Kingdom to show greater leniency by drawing up a Memorandum of Understanding that was harmless in content but formally compliant with Bucharest's proposal.[76]

Last but not least, the question of the delay in the negotiations with Rolls-Royce remained open. This was due both to the security considerations and also to the tedious trade negotiations with the Romanians. The first aspect will be discussed later, however it should be noted that in the absence of an agreement regarding certification, Romania would not be able to produce some of the key components for the Spey, and how this might objectively contribute to the efforts to maintain control over the transfer of technology. In the specific instance of the engine, this was a particularly thorny issue for the United States at CoCom.[77] As far as the business aspect was concerned, perhaps thinking that with the State visit just around the corner and with the BAe negotiations well on the way, it would be possible to lobby Rolls-Royce, Romania now said it wanted to buy only fourteen complete engines (instead of the fifty previously planned), along with a 30% discount and a counter-trade agreement covering all importation linked to subsequent Spey engines.[78]

Considering the tangle still to be unravelled, the United Kingdom's increasing unwillingness to make concessions and the conflicting attempts by each side to manipulate the imminent visit in order to force the other to capitulate, it is hardly surprising that even during the visit no definite conclusion to the negotiations was reached. The original plan to hold a solemn signing ceremony at the Banqueting Hall had to be

[76] UKNA, FV 14/93: note of a meeting between the Secretary of State for Trade and Mr. Pungan on 26.04.78; minute by Octon, 09.05.78; Secondé to FCO, tel. 200, 15.05.78; minute by Terry, 07.06.78.

[77] UKNA, BT 359/21: note of a meeting at DOI on 17.04.78; Barlow to Haynes, 15.05.78; minute by Terry, 14.06.78.

[78] UKNA, PREM 16/1838: state visit of President Ceaușescu – brief No. 1, n.d.; UKNA, FV 14/93: minute by Maynard, 23.05.78.

abandoned. All that happened was that during Ceauşescu's visit to Bristol on 15 June 1978, an instruction to proceed was finalised. This entailed, on the one hand, concession of the licence and, on the other, it committed the Romanians to pay around one million pounds in advance, to be covered with counter-trade in the course of 1978. This would allow BAC to get production under way. Still to be defined were the supply agreement and the purchasing agreement for the three complete aircraft on which the two sides had come to a compromise. The whole package would be implemented only after further bureaucratic procedures had been concluded by the end of September.[79]

Even the negotiations with Rolls-Royce, further complicated by security issues and the growing impatience of the company's management, resulted in a collaboration agreement that was explicitly linked to the one the Romanians had signed with BAe. It established the conditions for the production in stages in Romania – over 17 years – of 250 Spey 512-14DW engines, reaching 48% of the total value of the engine and for the training of Romanian technicians in the United Kingdom. Rolls-Royce also undertook to use its influence to gain an agreement on certifications and advantageous credit terms were offered to Bucharest. For its part, Romania agreed to purchase twenty-five (rather than fourteen) complete engines and to pay one million pounds sterling within two months of the contract coming into force, which would be by the end of September.[80] The Romanians also obtained an increase in the sterling component of the ECGD's financial cover (from 100 to 175 million pounds), on slightly better conditions and a Memorandum of Understanding on aerospace cooperation, which Boldur and Oprea had pressed for during the visit of the presidential delegation to London. This, however, was basically a rhetorical exercise expressing non-binding high hopes and bright expectations.[81]

All in all, the state visit was judged to be a success. The British press were critical of the welcome offered to the Romanian dictator but there were no embarrassing situations or unpleasant incidents, while the embassy in Bucharest reported clear signs of appreciation of the presidential couple.

---

79 UKNA, FV 14/93: minute by Chilton, 19.06.78; instruction to proceed, 15.06.78; protocol between, Bae, GAB and Tehnoimport, 16.06.78; "Romania Signs for BAC One-Elevens and R-R Spey Engines", *Trade and Industry*, 23.06.78, p. 673.

80 UKNA, PREM 16/1838: DTI to Mays, creda 108, 15.06.78; UKNA, BT 329/21: cooperation agreement between Rolls-Royce Ltd. and Tehnoimportexport, 15.06.78.

81 UKNA, FV 14/93: Secondé to Scott, 12.06.78; Cotterill to Stanciu, 15.06.78; note of a meeting between the Secretary of State for Trade and Mr. Oprea on 14.06.78; memorandum of understanding on aerospace cooperation, 15.06.78.

However, considering that the ROMBAC negotiations had been going on since 1976, it is hard not to interpret as a failure the fact that not even a state visit had been enough to bring them to a conclusion. Certainly, relations with Romania were still considered important, as shown by the fact that Callaghan asked Ceauşescu's opinion on how to react to China's interest in the Harrier, but there was no way that the magnitude of the ROMBAC project and the pomp and ceremony could conceal the absence of other, new prospects, not to mention the doubts on the actual possibilities of exporting a plane produced in Romania to CMEA countries or developing countries.[82]

5.3 Licensing the 1-11

As foreseen, failure to clinch the ROMBAC deal on the occasion of the state visit created a serious problem because, with the political propaganda value of the occasion no longer a reason for compromise, the Romanians started to repeat their requests for new concessions, perhaps on the assumption that with BAe's finances increasingly exposed, the British would be ready to comply. So, instead of closing by the end of summer as expected, the negotiations were protracted throughout 1978 on the guarantees that BAe was supposed to give to Romania regarding counter-trade. By the beginning of 1979, BAe had had enough and they sent a series of letters to both the Romanian and the British governments, declaring that they had reached their limit as regards financial negotiations and that unless the unresolved issues were settled quickly the ROMBAC project would be suspended. As a result, through diplomatic channels and the personal intervention of Keith, the Deputy Chairman of BAe, Allen Greenwood, requested a face-to-face meeting with Ceauşescu.[83]

The only person who still seemed to seriously consider the possibility of government intervention to help BAe to close the deal in the name of the political importance of bilateral relations was Ambassador Secondé, which aroused negative reactions in Whitehall. BAe saw the ROMBAC programme as important and profitable for the Weybridge-Bristol

82 UKNA, PREM 16/1838: record of PM's conversation with President Ceauşescu on 15.06.78; UKNA, FCO 28/3428: Bucharest to FCO, tel. 253, 16.06.78; UKNA, FCO 28/3429: Secondé to Owen, diplomatic report No. 184/78, 04.07.78; "Romanian One-Eleven Contract This Summer?", *Flight International*, 24.06.78, p. 1903.

83 UKNA, FV 14/158: Secondé to DOT, creda 103, 07.11.78; Owen to Bucharest, tel. 26, 25.01.79; Greenwood to Kaufman, 06.02.79; UKNA, FCO 28/3837: Greenwood to Palliser, 06.02.79; Keith to Ceauşescu, 15.02.79.

Division because of its links to the 1-11, but for the corporation as a whole its importance was less clear and in any case not sufficient to justify an agreement at all costs. In addition, though the Romanians showed an interest in the Aérospatiale N 262 (a small turboprop French plane with a military version) on the occasion of Valéry Giscard d'Estaing's visit to Romania in March 1979, there was no question of repeating the kind of competition that had been possible with the defunct 614. A similar project was technically unfeasible since, in order to stand as a viable alternative, the plane would have had to be completely redesigned (doubling the cabin space and replacing the turboprops with jet engines). Thus, in talks in Bucharest with Romanian ministers, the Deputy Permanent Undersecretary of the FCO, Reginald Hibbert, dismissed the idea, stressing instead the fact that an outrageous amount of time had been wasted on secondary issues and that the whole project was therefore at risk, with potentially disastrous effects on bilateral relations.[84]

The meeting with Ceauşescu on February 20 was cordial but inconclusive. The only positive thing that came out of it was that BAe was convinced by diplomatic staff to give up the idea of imposing an ultimatum, thus avoiding the eventuality of a definitive breakdown. Ceauşescu stated that contacts had already been made with other possible partners because, although the ROMBAC agreement was important, it was not necessarily binding. Apart from this thinly veiled threat, it was clear that what really mattered to the Romanians, even more than the guarantees on countertrade, was to agree on a formula that would block the escalation of prices. Actually, the British side had always stated clearly that, for a large long-term project such as the ROMBAC it would not be possible to fix a price as had happened in 1975, but with rising inflation in the United Kingdom, Ceauşescu repeated that it was now an internal problem that the British had to sort out for themselves. This idea was received with indifference in Whitehall because, unlike the situation of 1974-1975, now there was no employment issue at stake (it was a question of a total of 500 jobs in BAe which would anyway be absorbed into other divisions). The important thing was to avoid whetting the Eastern bloc countries' appetite for credit. BAe's board was also unimpressed and they decided that the preparatory activities for the ROMBAC would be downsized starting from March in order to reduce expenses. Considering that Ceauşescu's proposal

84 UKNA, FV 14/158: Secondé to DOT, creda 20, 06.02.79; minute by Chilton, 13.02.79; UKNA, FV 14/159: MacLeod to Battiscombe, 14.03.79; UKNA, FCO 28/3837: Secondé to DOT, creda 29, 15.02.79; UKNA, FCO 28/3827: speaking notes by Sir. Hibbert during discussion with Mr. Gliga on 14.02.79; meeting between Sir Hibbert and Mr. Andrei on 15.02.79.

to conclude the sale of the three 1-11s separately – enabling BAe to cover its expenses – was not followed up by the Romanian negotiators, in what was, actually, just a game, the only concrete result was a promise from the President to send a high level delegation to London at the earliest, to discuss the question of escalation.[85]

Significantly, this delegation was not led by the Minister for Foreign Trade, Cornel Burtică, but by the Minister of Finance, Paul Niculescu-Mizil, whose career until recent years had been in the education and propaganda apparatus of the party and the state. In 1976, he had moved to the economic sector as President of the committee for coordinating the production of consumer goods, becoming permanent Romanian representative to the CMEA in the 1980s. In the plethora of meetings in London between March 20 and 23, 1979, Niculescu-Mizil asked for concessions on virtually every aspect of the programme. From the industrial standpoint, it is important to note how, in the situation subsequent to the Islamic revolution in Iran (which brought about a serious recession in the aerospace market), the Romanian minister expressed doubts about the actual possibilities for exporting the ROMBAC. BAe, however, showed no inclination to consider a buyback. Secondly, there was the problem of certifications. In fact, a procedure had been outlined whereby CAA inspectors would monitor the construction and testing stages of the aircraft, so that, on the basis of a statement by BAe and Rolls-Royce that quality control had been satisfied, the CAA would attest that Romanian airworthiness certificate would be accepted in the case of ROMBAC imports to the United Kingdom. This procedure not only implied close cooperation between the CAA and the Romanian Department of Civil Aviation, all still to be put in place, but it also meant a significant increase in British support costs, because quality control and certification were usually down to the licensor. In addition, the problem would not necessarily be solved, as shown by the United States refusal to accept Romanian-manufactured Islanders, regardless of the fact that these had a CAA export certificate ratified by the United Kingdom.[86]

Financially, the crucial point remained the escalation clause. Though there was a possibility of tweaking the sterling credit component

---

85 UKNA, BT 359/11: Secondé to DOT, creda 33, 21.02.79; Secondé to DOT, creda 34, 21.02.79; Secondé to DOT, creda 35, 21.02.79; UKNA, FCO 28/3837: minute by Maynard, 23.02.79; minute by Scott, 06.03.79; UKNA, FV 14/158: minute by Terry, 26.02.79; minute by Terry, 28.02.79.

86 UKNA, FV 14/160: report on a visit to the UK by Mr. Niculescu on 20-23.03.79; UKNA, DR 8/27: Guilbert to Rain, 30.08.78; Chaplain to Terry, 22.08.78.

to cover the project (thus arriving at £200 million i.e. practically the entire programme, according to the latest cost estimates), the problem was Romania's unwillingness to accept an escalation greater than 7.5%, roughly the British inflation index in mid-1978, when the agreements were signed during a state visit.[87] Inflation had continued to rise in the second half of the year and then soared owing to the second oil shock. Callaghan was informed that the Romanians were demanding that the British government should pay the difference between the ceiling of 7.5% and BAe's final asking price, so the Prime Minister took charge personally – in spite of the delicate political moment in the United Kingdom[88] – to explain to Niculescu-Mizil how things really were:[89]

> […] the Prime Minister stressed that these decisions [in the field of civil aviation] had all been approached by the Government on the basis that any arrangement must be commercially viable and of commercial advantage to all the parties involved. This had also been the basis of his exchanges with President Ceauşescu on these matters. After careful review, the British Government had concluded that they had reached the limit so far as any further concessions to the Romanian position on the BAC 1-11 negotiations were concerned. The British Government had no interest in arrangements which ran counter to commercial considerations.[90]

This clarification had a salutary effect in the sense that it was now possible to save a deal that had been hanging in the balance. In fact, after more long-drawn-out negotiations, a new protocol was signed on April 5 that solved the problem of escalation: this was limited to only 7.5% per year for the three complete 1-11s. The purchasing contract was finally made truly independent from the rest of the package; base prices were updated to December 1977 and annual escalation was set at a maximum of 8.5%, a mechanism still favourable to the Romanians considering that average inflation in the United Kingdom between 1979 and 1981 was in double digits. To sweeten the pill, BAe undertook to ensure – once the ROMBAC had been finalised – the possibility of producing the Jetstream 3 and the Short 330 in Romania, and even to dig up the idea of selling the licence for

87 Historical inflation rates are available online at http://www.inflation.eu.

88 See Colin Hay, "Chronicles of a Death Foretold: the Winter of Discontent and Construction of the Crisis of British Keynesianism", *Parliamentary Affairs*, Vol. 63, No. 3 (2010), p. 446-470.

89 UKNA, FV 14/159: minute by Terry, 21.03.79; note of a meeting between Mr. Kaufman and Mr. Niculescu on 22.03.79; UKNA, FCO 28/3838: minute by Browning, 26.03.79.

90 UKNA, PREM 16/2223: Cartledge to Harris, 23.03.79.

the 748. These were planes that, in theory, should have replaced the defunct 614 (in the case of the 748) or the Aérospatiale N 262 (in the case of the Jetstream 3 and the Short 330); instead, in the global economic situation of the time, in particular that of Romania, they are more of a testimony to the levels of rigidity and megalomania inherent to Romanian planning.[91]

> Apparently what the Romanians envisage is taking over complete production of 748s and leaving BAe simply the task of selling them (to a parallel to the Islander situation). They appear to be completely unrealistic about the implications and seem to think that they could be in full production in a year or two. This attitude is tending to dampen the enthusiasm of BAe [...].[92]

In April 1979, it was thus possible to finally resolve the last outstanding issues, namely credit conditions offered by ECGD. On the one hand, the British offered an interest rate of 7.75% over ten years, while the Romanians demanded 6.5% over 15 years. For Bucharest, the general climate was unfavourable, because at the time the question of non-tariff barriers in the civil aviation market, also in the form of interest rates on loans, was at the centre of a heated debate between the Americans and the Europeans at the Tokyo Round.[93] Apart from this, however, the British were growing increasingly weary, which meant that they were only willing to make those marginal concessions necessary to ensure that, should the negotiations fail, the responsibility would not be attributed to London. It was thus agreed that the interest rate would remain unchanged at 7.75%, but the period of payment would be reformulated to 15 years from the date the contract came into effect.[94]

On May 3, 1979, the Romanians finally accepted the credit conditions and all contracts with BAe were signed by the end of the month in time to be made public on June 8, shortly before the Paris Air Show. On June 20, contracts were signed with Rolls-Royce for the licensed production of Spey.[95] So, after tedious negotiations that had lasted around four years,

---

91 UKNA, FV 14/159: Greenwood to Avram, 04.04.79; protocol signed on 05.04.79; UKNA, FV 14/160: Brown to DOT, creda 51, 08.06.79; UKNA, FCO 28/3838: Secondé to Dearing, 04.04.79.

92 UKNA, FV 14/160: minute by Chilton, 02.10.79.

93 Hayward, *International Collaboration*, *op. cit.*, p. 170-175.

94 UKNA, FV 14/160: note of a meeting in Bucharest on 10.04.79; Cotterill to Widdup, 18.04.79; Widdup to Cotterill, 25.04.79; UKNA, FCO 28/3838: Secondé to FCO, tel. 87, 18.04.79; Kaufman to Niculescu, 30.04.79.

95 UKNA, FV 14/160: Secondé to DOI, Indus 5, 03.05.79; minute by Chilton, 07.06.79; minute by Miller, 20.06.79; UKNA, FCO 28/3838: minute by Mawson, 04.06.79.

during which the aircraft had become increasingly obsolescent, and at the start of a ruthlessly recessive phase for civil aviation, Romania set about building a large airliner.

It should be made clear that even at this point the negotiations were not definitively closed. Even for civil aviation projects, export controls played an important part, and the negotiations happened to occur at a time when CoCom's activities were coming under fresh scrutiny. In the United States, the Bucy Report of February 1976 had established that what should be controlled and possibly banned was not a specific product, but the technology that the product embodied. This approach, though open to different interpretations, highlighted the importance of maintaining technological superiority over the USSR. In practice, nothing much changed as far as CoCom's position was concerned, not least because of European scepticism and hostility regarding the potential effects on East-West exchanges and technology transfers within the Western world.[96]

In this changing climate, the vicissitudes and sluggishness of the ROMBAC negotiations provoked worry and annoyance, especially in view of Romania's increasing role in the future production of the aircraft. BAe had kept the design authority so the problem of cooperation in design technology did not arise and it was assumed that Romania would never be able to bridge the technological gap. The airframe posed no problems for CoCom, provided the machine tools required for manufacturing were not too advanced (basically, excluding those that were computer numerical controlled.) The so-called 'bought-out items' were a different matter (for example the undercarriage and the avionic systems, which BAe obtained from other suppliers), which mostly came under a different heading on the CoCom embargo list. In any case, 'bought-out items' were to be supplied in quantities that were strictly necessary for the realisation of the project. The real problem was the production of the Spey under licence; according to Whitehall officials CoCom would be overwhelmingly opposed to this. They hoped that if they could manage to get the go-ahead for the fuselage, it would be possible to obtain approval for the engines, but when CoCom did approve the ROMBAC project on April 13, 1978, they specified that the only information relating the Spey that could be passed on to the Romanians was that necessary for fitting the engines on to the plane.[97]

---

96 Mastanduno, *Economic Containment, op. cit.*, p. 186-219.

97 UKNA, BT 359/9: note of a meeting held on 08.07.76; Haynes to Williams, 21.03.77; UKNA, BT 359/10: minute by Gunns, 23.05.77; UKNA, BT 359/11: CoCom (78) 89, 14.02.78; CoCom (78) 89.4, 13.04.78.

Seeing that the Romanians were already producing the Viper under licence and collaboration on the military version of the Spey was not allowed, the limitations imposed were decidedly excessive and might have been relaxed at an opportune moment. However, the DOI maintained that this would not be possible before the state visit and that in any case it was inadvisable to risk blocking negotiations on the airframe by telling Bucharest that for the time being it was not possible to collaborate on the civil version of the Spey. Rolls-Royce would have to drag things out for as long as possible and submit any agreement reached for government approval.[98]

As we have seen, CoCom's indications on the approval of the ROMBAC project did not prevent the signing of the framework agreement for cooperation on the Spey 512, but a constraint remained on the metallurgical know-how needed to produce castings and forgings for the engine. The issue was particularly sensitive because Romania had shown interest in metallurgical techniques since 1968 and at the time of the licence for the Viper, in 1972, the Romanians had been granted the rights to produce the castings and forgings of the engine. Actually, this meant nothing, firstly because the technical documentation produced by Rolls-Royce was incomplete (the data relating to work that was not Rolls Royce's direct responsibility was missing, for example the forgings produced by High Duty Alloys – a subsidiary of HSA) and secondly because the Romanians lacked the basic skills to use it. With the ROMBAC negotiations under way, in November 1976, Bucharest asked to be given the entire technological know-how on metals used in aerospace and a consortium of companies, including Rolls-Royce, was called upon to look into the request. Knowing that technology was a problem for CoCom, Rolls-Royce strove to eliminate castings and forgings from the cooperation agreement on the Spey despite vehement protests from the Romanians. Yet at the meeting of June 14, 1978, Oprea noted that there was no point in talking about 'cooperation' on the Spey if the British were not prepared to transfer the know-how on something which constituted more than half of the engine. Dell pointed out that the United Kingdom had constraints to respect, but that it would certainly provide Romania with the necessary materials to manufacture the engine.[99]

So the castings and forgings were the subject of a special protocol on metallurgy for use in aviation and the British consortium undertook only

98 UKNA, BT 359/11: Chilton to Haynes, 21.04.78.

99 UKNA, BT 359/21: minute by Maynard, 02.06.78; minute by Terry, 16.06.78; UKNA, BT 359/23: Moir to Mann, 21.05.81.

to make an offer limited to the basic technology for the production of alloys and in pre-set casting and wrought components. Any agreement would have to be approved by the British Government. In a separate letter, Rolls-Royce stated that it could not guarantee the transfer of all metallurgical know-how necessary for the production of the Spey.[100]

In October 1978, the Consortium presented an offer for a total cost of 35 million pounds. The magnitude of the figure pushed the Romanians to focus on castings and forgings, and to shelve the basic techniques for the production of alloys and special steels, but in this more restricted area, the know-how for the manufacture of turbine blades and core manufacture technology was not included because of British considerations relating to CoCom – especially the fear that the technology would then be surreptitiously passed on to the Soviets. These elements, like all the castings, would be sold directly to the Romanians by Rolls-Royce.[101] It is difficult to understand what possible interest the Soviets could have in the technology embodied by a rather old engine; however, it should be noted that even with these strong restrictions on the transfer of know-how, the negotiations with CoCom were not simple and it was only in October 1979 that American and German reservations were withdrawn.[102]

Romania did not give up on the castings and forgings, however. In April 1981, Prime Minister Verdeţ visited the United Kingdom, and whereas the British Government wanted to discuss the big political issues of the moment such as Afghanistan, the Polish crisis, and the renewed tension between the two blocs, the Romanians were more anxious to talk about the bad state of the bilateral trade balance and gain London's backing in order to secure more export opportunities in the EEC. In particular, however, Verdeţ told Thatcher and her ministers that the limitations imposed on the metallurgical industries were obstructing the implementation of the ROMBAC project. The Secretary of State for Industry, Keith Joseph, promised that the matter would be reviewed and indeed, this time, the Romanians were more fortunate.[103]

---

[100] UKNA, PREM 16/1838: DTI to Mays, creda 108, 15.06.78; UKNA, FV 14/93: protocol of 09.06.78.

[101] UKNA, BT 359/22: minute by Todd, 03.01.79; Henry to Miller, 08.01.79; minute by Miller, 01.02.79.

[102] UKNA, BT 359/22: Miller to Moir, 17.08.79; Miller to Henry, 19.09.79; Fisher to Robson, 18.10.79.

[103] UKNA, PREM 19/580: minute by Armstrong, 17.03.81; note of a meeting between the PM and Mr. Verdeţ on 13.04.81; UKNA, BT 359/23: Mann to Nickel, 30.04.81.

In the wake of the Government's recent refusal to negotiate the licence for the Spey 202 with Bucharest, and faced with the shutdown of the 1-11 production line in the United Kingdom, Rolls-Royce wanted to sell the technology in question to Romania in order to make a profit as long as it still had some commercial value. With the development of more and more advanced engines, this technology would soon be obsolete, and from 1985 China would be free to dispose without restrictions of the technological know-how obtained with the licence for the Spey 202. Rolls-Royce, therefore, prepared an offer split into three packages: the first included the basic information required to make full use of the know-how already received with the licences for the Viper and the Spey 512; the second and third packages concerned the castings and forgings, whose intellectual property belonged to Rolls-Royce, along with the skills that the Romanians would need to produce the components of the 512 that they were already entitled to manufacture under the terms of the licence.[104]

Finally, in September 1981, the British Government authorised Rolls-Royce to enter into negotiations with the Romanians. There were still some restrictive conditions, for example the exclusion of the technology related to directionally solidified casting and mono-crystal casting – an important part of the package. However, as noted by Stephen Bickel, an official in the Commercial Relations and Exports Division of the DOT, with Romania about to take draconian measures to repay its foreign debt, the problem would be mostly, if not exclusively, commercial rather than technological.[105] In actual fact, Anglo-Romanian aerospace cooperation was soon to come to an end.

[104] UKNA, BT 359/23: Transfer of Rolls-Royce Ltd. Casting and Forging Technology to Romania, 05.81.

[105] UKNA, BT 359/23: Wilding to Bickel, 17.07.81; Bickel to Moir, 15.09.81.

Chapter 6

Epilogue: Re-engining the 1-11

Andrei's memoirs devote a large amount of space to significant aspects of Romania's political economy in the 1980s: namely, the obsession with export goals, the quality problems of manufactured goods, the asphyxiating and dysfunctional centralisation, the aversion of an ageing Ceauşescu to electronics and robotics – considered possible causes of social problems, such as unemployment or excessive leisure time.[1] The most interesting thing, however, is what is not immediately apparent, that is, the fate of that aviation industry in which the Romanian regime had invested so much and which seemed to be reduced to the supply of helicopters and spare parts to Sudan and Angola.[2] In fact, the 1980s tolled the death knell for the ambitious projects of the previous decades. It was during these years that two key features faded from the scene: the neo-Keynesian attention to supporting a branch of industry considered crucial in terms of innovation and employment – if not of prestige – in Britain and Romania's diminishing political relevance as the maverick of the Soviet bloc. A neo-liberal approach and a progressive marginalisation would result, after Ceauşescu's death and the fall of the Communist regime, in the demise of Anglo-Romanian cooperation in the aviation industry.

6.1 The ROMBAC Programme in '*Epoca de Aur*'

From 1978, the growth of the Romanian economy began to decelerate. Of course, its figures were still high compared to Western and Soviet bloc countries, even if statistics should be taken with a pinch of salt. However, investments – which historically had always been kept at very high levels – began to decline, especially in the capital goods sector.[3] Priority was

[1] Ştefan Andrei, *I se spunea Machiavelli. Ştefan Andrei în dialog cu Lavinia Betea*, Bucharest, Adevărul, 2011, p. 396-413.

[2] *Ibid.*, p. 488-499.

[3] Lloyds, f. 9640, HO/D/Boa/Age/50: note on Romania by the Economics Dept, July 1980.

given to the completion of ongoing projects and the modernisation of existing capacities, with a focus on those low-intensity raw materials and energy which, nevertheless, could significantly contribute to exports while launching a New Economic Mechanism that – unlike the Hungarian experiment of the same name – concealed, under the guise of corporate self-management, an increasingly rigid centralist plan based on material indicators.[4]

The difficulties of the economy stemmed from the very structure that it had assumed during the years of forced industrialisation. This structure had generated enormous productive capacities in sectors such as petrochemicals, unhappily exposed to the effects of oil shocks and of the global slowdown in growth. All this occurred as the rise in interest rates transformed loans, used to build industries that were at that moment struggling to find outlets or even just to be used with low profitability, into a dangerous burden for Romania. For example, the trade balance in the oil sector had gone from a small surplus in 1976 to nearly $2 billion in deficit in 1976 (when Romania's crude oil production had hit a peak.) This situation, together with reckless financial management, resulted in a sharp deterioration of the external payments position and in a gradual shift of Romanian exchanges towards the CMEA, in order to obtain raw materials from the Soviet Union on favourable terms. Although in the 1980s, foreign trade was still generated by 33.7% with the CMEA countries, by 35.5% with Western and 30.8% with developing countries, by 1988 the respective percentages were 47.8%, 27.3% and 24.9%.[5]

Moreover, the Polish crisis of 1980-1981 clearly showed the wider implications of an economic policy that had failed to control the trade deficit and the growth of foreign debt. Even though Whitehall officials correctly noted how in Romania essential operative conditions like Poland's (a solid and well-organised dissident movement, a tradition of workers' protests and an alternative point of reference, such as the Roman Catholic Church) were lacking,[6] the situation contributed to generating an atmosphere that further damaged the creditworthiness of the country, – something that the Romanian leadership closely followed.[7] Ceaușescu,

---

4 Swain – Swain *Eastern Europe Since 1945*, *op. cit.*, p. 175; Wiener – Slater, *East-West Technology Transfer*, *op. cit.*, p. 82-85.

5 Lloyds, f. 9642, HO/D/Boa/Age/51: note on Romania by the Economics Dept, September 1981; Foreign trade percentages are reported in Murgescu, *România şi Europa*, *op. cit.*, p. 358.

6 UKNA, CAB 133/522: MV(81)1(b), Romanian internal situation, 09.04.81.

7 Petre Opriş, *Criza poloneză de la începutul anilor 80. Reacţia conducerii Partidului Communist Român*, Ploieşti, Petroleum-Gas University Publishing House, 2008.

who had lived a genuine moment of glory by denouncing the invasion of Czechoslovakia in 1968, now strongly advocated an intervention that would put an end to the experience of a free trade union within a Communist country:

> The Polish Government must put its own house in order. Otherwise others would do it for them and could not be blamed for doing so. [...] The Dubček experiment had been conducted within the framework of Socialism and, whatever one might think of it, was perfectly compatible with a socialist system. What was going on in Poland was not a variant of Socialism but a challenge to the system as such.[8]

Ceauşescu, while recognising the adverse effects of certain circumstances, such as the oil crisis and rising interest rates, attributed the responsibility of the economic difficulties to the mistakes and shortcomings of the Communist leadership in the implementation of five-year plans, whose mechanism, along with his own leadership, thus escaped critical analysis.[9] In this way, for example, Foreign Trade Minister Burtică, who had played a significant role in Anglo-Romanian negotiations in 1969-1972, was ousted in May 1982 on charges of bribery and thievery – accusations alleged by Burtică to have been purposely set up by the Securitate, which, during the 1980s, became increasingly involved in currency operations of dubious legality in competition with economic units.[10] In addition, Law No. 12 of December 1980 established that international trade had to rely on the exchange of goods as much as possible: strong currency payments were to be used only if the importing economic unit was able to balance out expenses through exports.[11]

In the 1980s the top priority of the regime was to wipe out foreign debts so as to protect Romania from those external influences that were felt in other Eastern European Communist countries. This was achieved by boosting the volume of exports and slashing imports, by dramatically

---

8 Papers of Julian Amery, Churchill College, Cambridge (hereafter AMEJ), 10/1/45: note of a conversation with President Ceauşescu on 16.10.80.

9 Nicolae Ceauşescu, *Cuvîntare la consfătuirea de lucru de la CC al PCR din 29-30 mai 1980*, in *Id.*, *România pe drumul construirii societății socialiste multilateral dezvoltate*, Vol. 20, Bucharest, Ed. Politică, 1981, p. 69-108.

10 Rodica Chelaru, *Culpe care nu se uită. Convorbiri cu Cornel Burtică*, Bucharest, Curtea Veche, 2001, p. 206-211; Florian Banu, „Capitaliştii avant la lettre: Securitatea şi operațiunile valutare speciale din anii 80", *Caietele CNSAS*, Vol. IV, No. 1-2 (2011), p. 109-136.

11 Legea No. 12/1980 din 19 decembrie 1980 pentru întărirea autoconducerii muncitoreşti şi autogestiunii economico-financiare şi valutare în activitatea de comerț exterior şi cooperare economică internațională, available at http://lege5.ro/.

reducing consumption on the part of the people, who were thus called upon to endure huge sacrifices. In his report to the RCP National Conference in December 1982, Ceauşescu – in formal homage to business as usual – emphasised this priority unequivocally:

> Particular attention has been paid to achieve an active trade balance, acting with determination to reduce imports and increase exports. [...] Romania stands firmly against any measure of restriction or protectionism in international economic exchanges. We hold that, in the current conditions of economic crisis, it is necessary not to restrict, but to intensify and develop exchanges, to collaborate and cooperate in production. On this basis, in 1983 and in the years to come, we will intensify foreign trade. As regards the balance of payments, we will pay particular attention to the continuous reduction of imports and to the growth of exports, in order to create the necessary conditions for the gradual extinction of external debt. We believe that we will reduce at least half the level of foreign debt by the end of 1985 and, in the next 2-3 years of the 1986-1990 five-year period, to completely liquidate it. In the conduct of foreign trade activities, emphasis will be put on the realisation of balanced trade, not allowing any use of credits to increase external debt. Insisting on the achievement of balanced economic exchanges, in order to compensate for imports and exports, we do not intend to limit economic relations and trade; on the contrary, we believe that this can create the conditions for expanding and developing our international economic exchanges.[12]

It is a fact that Romania's creditworthiness at the beginning of the 1980s was a problem. In 1982, the ECGD offered only short-term credits backed up by an irrevocable confirmed letter of credit, though very reluctantly. For example, when Corsarc Engineering Ltd. – located in an area with one of the highest unemployment levels in the United Kingdom – asked for support to export the special furnaces to be used in the Spey production line in Romania, the request was turned down, even if the maintenance and possible recovery of employment levels depended on that order.[13] Although an austerity policy was necessary, it was Ceauşescu's choice to pay off the entire foreign debt over a short period of time (the debt was actually extinguished by the spring of 1989.) In addition to transforming Romania into the 'Ethiopia

---

[12] Nicolae Ceauşescu, *Raport la Conferința Națională a Partidului Comunist Român, 16-18 decembrie 1982*, Bucharest, Ed. Politică, 1982.

[13] UKNA, T 437/76: Newton to Adams, 27.01.82; Blackburn to Adams, 12.02.82; Adams to Blackburn, 22.02.82.

of Europe',[14] this choice prevented foreign trade from becoming a stimulus to the whole economy, which was becoming more and more autarchic.[15]

Meanwhile, there was talk of intensifying scientific and technological cooperation and of a renewed effort to assimilate new production technologies. As has been pointed out, the reason why such measures failed was that the targets were the same as twenty years before, without obviously proposing concrete remedies for past mistakes.[16] Even Ceauşescu, in his report to the 13th Congress of the RCP in 1984, stated that 'Romanian science' ought to ensure that, by the end of 1990, 95% of industrial products would be sold on the international market and 2-5% would be in the top quality range.[17] The real break with the past, however, was the new importance given to cooperation within the CMEA, which Romania had previously strongly opposed, but was now forced to support:

> Romania will act with the utmost determination to implement the agreements to improve the work of the CMEA, the development of economic exchanges and cooperation in production. We will be actively involved in the coordination of plans, in the specialisation of production and in common scientific and technical research activity, so that the CMEA can play an increasingly important role in the economic and social development of its member countries, in the building of Socialism and Communism, in enhancing the well-being of our peoples. (Long thunderous applause)[18]

Writing to Margaret Thatcher in 1983, former Conservative aviation minister Julian Amery noted that the Romanian president looked physically weakened and the country had fallen prey to a palpable malaise:

> Both Robert [Cranborne] and I felt that there was an indefinable atmosphere of crisis which they were doing their best to conceal by a show of serenity. We thought this probably concerned East/West relations. But from the little we saw of goods in the shops and queues at petrol pumps, several hundred cars long, the Romanian economy is clearly in difficulty.[19]

14 Liviu Ţăranu (ed.), *Românii in „epoca de aur". Corespondență din anii 80*, Târgovişte, Editura Cetatea de Scaun, 2012, p. 264-265.

15 Aldcroft, *Economic Change in Eastern Europe*, *op. cit.*, p. 170-171.

16 Wiener – Slater, *East-West Technology Transfer*, *op. cit.*, p. 85.

17 Nicolae Ceauşescu, *Raport la cel de-al XIII-lea Congres al Partidului Comunist Român*, Bucharest, Ed. Politică, 1984, p. 23.

18 *Ibid.*, p. 25.

19 AMEJ 10/1/45: Amery to Thatcher, 28.01.83.

Various testimonies remind us of how toxic the air had become in a country made grey and shabby, whimsically driven by a dictator who had begun to believe in the propaganda incessantly fuelled by his increasingly grotesque personality cult.[20] Mikhail Gorbachev told the Soviet Politburo of the sense of oppression he experienced during his official visit in 1987, the clash between the triumphalism of proclamations and the miserable living conditions of people, as well as the incredible audacity and presumptuousness of the ageing Ceauşescu.[21]

Somewhat paradoxically, the arrival of Gorbachev and the beginning of Perestroika closed the space for manoeuver and dealt the final blow to Ceauşescu's international image – and not just because of the contempt that the Soviet General Secretary never failed to demonstrate for a regime that he considered conceivable in North Korea, but not in civilised Europe.[22] Precisely when the Soviet Union appeared to be convincingly on the road to emerging from the Cold War and implementing internal reform, Ceauşescu repeatedly condemned the tenets of Perestroika:

> But the general problems of development impose a single policy, the state leadership, on the basis of the single national economic and social development plan. [...] We understand the role of economic laws, but it is difficult to understand how, in that spirit, the regulation of the problems of economic and social development can be left to the law of supply and demand, to market laws. It is difficult to guarantee progress through so-called market Socialism. [...] In these circumstances, the state is not only unable to renounce its function, but it has to face it in better conditions, in a democratic way. Undoubtedly, state functions are transformed, perfected, but they do not disappear nor can they ever disappear.[23]

---

20 See Anneli Ute Gabanyi, *The Ceausescu Cult. Propaganda and Power Policy in Communist Romania*, Bucharest, Ed. Fudaţiei Culturale Române, 2000.

21 Svetlana Savranskaya, Thomas Blanton and Vladislav Zubok (eds.), *Masterpieces of History. The Peaceful End of the Cold War in Europe, 1989*, Budapest – New York, CEU Press, 2010, doc. 13, p. 253-254.

22 Gorbachev made this remark in a conversation with Helmut Kohl in mid 1989, see *Ibid.*, doc. 67, p. 476-478.

23 Nicolae Ceauşescu, *Raport cu privire la stadiul dezvoltării forţelor de producţie, al societăţii socialiste în general, al relaţiilor de producţie şi sociale, al democraţiei muncitoreşti-revoluţionare, al perfecţionării conducerii, pe baza principiilor autoconducerii şi autogestiunii al perfecţionării şi continuării procesului revoluţionar în noua etapă de făurire a socităţii socialiste multilateral dezvoltate şi de înaintare a României spre comunism*, 14.12.87, in *Id.*, *România pe drumul construirii societăţii socialiste multilateral dezvoltate*, Vol. 31, Bucharest, Ed. Politică, 1989.

Now that relations between the USSR and the West were radically changing, the political parabola of Ceauşescu – once the maverick of the Communist world – was set to end in the international condemnation of human rights abuses by his sick regime, which had interfered grossly in all aspects of the private life of individuals, including reproduction and where to live.[24] For example, the process of 'systematisation,' that is, the relocation of peasants from traditional villages to urban conglomerations, aroused vehement protests around the world.[25] In 1989, the Prince of Wales, Charles, publicly stated:

> We in this country are painfully aware of the trauma caused by uprooting traditional communities at the behest of 'benevolent,' know-all planners. We hope that we have learned something from such an experience. That process should have made us, therefore, all the more sensitive to the awful spectre of an entire society – not just certain districts – losing its roots and its ancient communities, which is what is happening today in a corner of Eastern Europe, in a country called Romania. There, President Ceauşescu has embarked on the wholesale destruction of his country's cultural and human heritage. […] The object, which is very interesting, is to reshape the nation's identity, to create a new type of person, utterly subordinate to its dreams. To achieve this, President Ceauşescu has set about destroying the cities and villages of his country and replacing them with blocks of flats which are a repetition of failed 1960s social engineering, mixed with the atmosphere of George Orwell's 1984.[26]

This involution also had an impact on the development of the Romanian aeronautical industry, which, in 1979, was still mentioned as one of the priorities in the country's development.[27] Likewise, such an atmosphere did not encourage foreign tourists to visit Romania: after the efforts and investments of previous decades, arrivals dwindled after 1981.

---

24 See David Funderburk, "Relations Between the United States and Romania During the First Half of the 1980s", in Quinlan (ed.), *The United States and Romania*, *op. cit.*, p. 147-159 and *Id.*, *Pinstripes and Reds. An American Ambassador Caught Between the State Department and the Romanian Communists, 1981-1985*, Washington, D.C., Selous Foundation Press, 1987; Gail Klingman, *The Politics of Duplicity. Controlling Reproduction in Ceausescu's Romania*, Berkeley, University of California Press, 1998.

25 Dennis Deletant, *Ceauşescu and the Securitate. Coercion and Dissent in Romania 1965-1989*, London, Hurst, 1995, p. 294-319.

26 Speech by H.R.H. The Prince of Wales at the opening of the Build A Better Britain Exhibition, Islington, London, The Business Design Centre, 27.04.89 available at http://www.princeofwales.gov.uk/media/speeches/speech-hrh-the-prince-of-wales-the-opening-of-the-build-better-britain-exhibition-the.

27 Nicolae Ceauşescu, *Raport la cel de-al XII-lea Congres al Partidului Comunist Român*, Bucharest, Ed. Politică, 1979, p. 29.

While the international economic crisis of the early 1980s was already affecting those tourists in search of economical solutions – the very ones that Romania was targeting – the regime adopted a suspicious and hostile attitude towards visiting foreigners, who were subjected to a series of vexing measures and to the intrusive checks of the Securitate, as well as having to settle for services below international standards in a country where everything was closed by 10 p.m.[28] Papers belonging to the Securitate document of dirt, means of transport in deplorable conditions, drunken staff, bribes extorted in exchange for decent meals, etc.[29] Regardless of whether it was a deliberate choice or a mere negligence,[30] it was certainly a manifestation of incompetence,[31] especially considering that, twenty years before, TAROM's endowment with a modern fleet had also been justified as a way of obtaining a strong currency through the expansion of tourist flows from Western European countries.

Meanwhile, Britain was also undergoing profound changes under Thatcher. This is not the appropriate place to dwell on the never-ending myth of a supply-driven economic policy aimed at restoring value to savings, which successfully reshaped the Keynesian precepts on employment and income distribution, and which finally put an end to the country's decline by returning to fundamental market laws.[32] Research, after all, has highlighted how, while not in power, the Conservatives had said very little about possible monetary control mechanisms. It has also shown how emphasising the issue of inflation as if it were a long-term illness of the national economy was simply a way of pandering to Thatcher's intimate beliefs and gaining rewards on the electoral level; and what is more, it allowed unemployment to be put on the backburner.[33] In this scenario, the decline and fall of corporatism and Keynesianism during the

---

[28] See David Turnock, *The Economy of East-Central Europe, 1815-1989*, London, Routledge, 2004, p. 405-408; Nicolae Costantinescu (ed.), *Istoria economică a României*, Vol. II, Bucharest, Ed. Economică, 2000, p. 312-320 and Derek Hall (ed.), *Tourism and Economic Development*, *op. cit.*, p. 101-102.

[29] Florian Banu (ed.), *„Amorsarea" Revoluției. România anilor '80 văzută prin ochii Securității*, Târgoviște, Editura Cetatea de Scaun, 2012, p. 101-103, 106-108, 304-309.

[30] See Ilie Rotariu, *Globalizare și turism: cazul României*, Sibiu, Continent, 2004, p. 218-219.

[31] Murgescu, *România și Europa*, *op. cit.*, p. 380.

[32] See Cosimo Magazzino, *La politica economica di Margaret Thatcher*, Milano, Angeli, 2010, p. 57-78.

[33] Jim Tomlinson, "Thatcher, Monetarism and the Politics of Inflation", in Ben Jackson and Robert Saunders (eds.), *Making Thatchers Britain*, Cambridge, Cambridge University Press, 2012, p. 62-77.

'Winter of Discontent' was not precipitated by intrinsic contradictions, but by an purposely-orchestrated political discourse, while the ostentatious monetarist refusal of an income policy was aimed at isolating central policy-making from the influence of various social interests.[34]

Some critics have underlined the costs and the chance nature of the outcomes of the fight against inflation, determined rather by high and unexpected level of interest on the exchange rate than by a medium-term financial strategy based on questionable theoretical assumptions and on misleading data relating to money supply.[35] Of course, the first Thatcher government marked the end of any possible reincarnation of Edwardian paternalism and it contributed in a decisive way to transforming the face of the British economy, dramatically accelerating the de-industrialisation process – particularly at a rate which seriously compromised exports for the manufacturing industry – and implementing a series of liberalisations that, coupled with the collapse of technological cost barriers, led to the so-called 'Big Bang' of the London Stock Exchange. Long story short, the British economy experienced a shift towards outsourcing, which was in turn largely responsible for the increase in productivity indexes, while the characteristics of economic welfare were redefined by de-industrialisation far more than by monetarist experiments.[36]

De-industrialisation was accompanied by the privatisation strategy, which was central to Thatcher's second mandate. Thus, after the first experiments with BP, British Aerospace and the British Sugar Corporation – all privatised by 1981 – communications and gas public monopolies were marketed, and so were British Airways, Rolls-Royce and British Airports Authority. Again, beyond the theoretical debates on the possible effects of

---

34 Hay, "Chronicles of a Death Foretold, *op. cit.*, p. 446-470; Jim Bulpitt, "The Discipline of the New Democracy. Mrs. Thatcher's Domestic Statecraft", *Political Studies*, Vol. 34, No. 1 (1986), p. 19-39.

35 David Cobham, *The Making of Monetary Policy in the UK, 1975-2000*, Wiley, Chichester, 2002, p. 137-152; Duncan Needham, *UK Monetary Policy from Devaluation to Thatcher, 1967-1982*, Basingstoke, Palgrave, 2014, p. 67-82.

36 Richard Vinen, *Thatcher's Britain. The Politics and Social Upheaval of the 1980s*, London, Simon&Schuster, 2009, p. 101-133; Eric Evans, *Thatcher and Thatcherism*, London, Routledge, 2013, p. 31-34; Jacques Leruez, *Thatcher: la dame de fer*, Bruxelles, Andre Versaille, 2012, p. 123-124; Charles Goodhart, "The Economics of 'Big Bang'", in Ranald Michie (ed.), *The Development of London as a Financial Centre*, Vol. IV: *1945-2000*, London, Tauris, 2000, p. 261-277; Jim Tomlinson, "De-industrialisation Not Decline: A New Meta-narrative for Post-war British History", *Twentieth Century British History*, Vol. 27, No. 1 (2016), p. 76-99.

reorganisation and the increase in productivity induced by privatisations,[37] the responsibility for industrial investments had by now been shifted to the private sector. On the one hand, this meant that businesses would no longer be subjected to constraints imposed by the Treasury, weighing on taxpayers, but instead they could benefit from the growing globalisation of capital markets. On the other hand, the government could make money and finally get rid of Labour's social contract policy to favour the Trade Unions, which were already seriously debilitated by the industrial employment crisis. Thus, at the end of the 1980s, the state spent about 13% more than ten years earlier, but in relation to the entire economy, the weight of this had actually been offset by the growth of the private sector. The composition of public spending had also changed, with the virtual disappearance of industrial intervention policies (the DTI budget fell by 38%).[38]

Although these changes could not affect the way Anglo-Romanian aeronautical relations were managed, Romania's growing marginality on the British diplomatic front should still be considered. Thatcher's arrival at Downing Street did not substantially change international relations.[39] Known as the 'Iron Lady,' a nickname coined for her by the Soviets, the new Prime Minister continued to view foreign policy from the perspective of the Cold War. She was convinced of the repressive nature of Communism – in her view teleologically destined to perish – and deeply critical of Détente in the 1970s. However, after 1983, her tones softened and, in February 1984, she even visited János Kádár's Hungary, the first of several trips to Eastern Europe. The aim was, basically, to implement a policy of differentiation, favourable to those countries willing to show tentative openness and to exploit the new possibilities offered by Perestroika. Of course, it was assumed that the situation in Eastern Europe would not change radically any time soon and that it was, therefore, necessary to sustain the spirit of freedom of the people oppressed by Communism, by preaching the virtues of the free market.[40]

---

37 See Andrew Gamble, "Privatization, Thatcherism, and the British State", *Journal of Law and Society*, Vol. 16, No. 1 (1988), p. 1-20.

38 Graham Steward, *Bang! A History of Britain in the 1980s*, London, Atlantic Books, 2013, p. 179-184, 380-385; Evans, *Thatcher and Thatcherism, op. cit.*, p. 34-38; Leruez, *Thatcher: la dame de fer, op. cit.*, p. 126-130.

39 Vinen, "Thatcherism and the Cold War", in Jackson – Saunders, *Making Thatchers Britain, op. cit.*, p. 199-217; Paul Sharp, *Thatcher's Diplomacy. The Revival of British Foreign Policy*, Basingstoke, Palgrave, 1997, p. 183-211.

40 Evans, *Thatcher and Thatcherism, op. cit.*, p. 114-117; Leruez, *Thatcher: la dame de fer, op. cit.*, p. 165-166.

It was countries such Hungary or the Poland of Solidarność that stood to benefit from this approach, certainly not Romania, subjected as it was to a grim and sometime grotesque dictatorship.[41] In the 1980s, therefore, this differentiation policy, which in the past had long played in favour of Bucharest, began to have the opposite effect. The transformations of the British economy were pushing towards the liquidation of commitments with Romania, since Whitehall was no longer willing to provide financial support to industries – except in certain cases[42] – and businesses were urged to make choices based on profit in a country in the throes of a seemingly endless economic crisis. It is no coincidence that, in her memoirs, Thatcher refers to Romania only once, when she mentions Ceauşescu's ignominious demise.[43]

At the beginning of the 1980s, relations between BAe and the British government were rather strained for what concerned the 1-11. Even during the nationalisation era, there had been talk of putting an end to the original launching aid and to the subsequent underwriting agreement, as these were considered inappropriate forms of financial support for a public enterprise. In April 1979, BAe asked the government to renounce its claims relating to the licence fee – even though it was entitled to this under the launching aid agreement – in order to conclude the ROMBAC negotiations with Romania. Whitehall saw this as a good opportunity to finally settle the issue.[44]

The rapid turnover of officials at the DTI and at the Procurement Executive of MOD (which, after the abrogation of the Ministry of Aviation Supply in 1971 had served as the agent for the DTI in administering civil aviation support contracts) and the delay in proceeding with the sale of aircraft No. 9 covered by the underwriting agreement (only at the beginning of 1984) meant that contacts became intermittent, with the parties being increasingly unable to take control of the accounting of the money involved. In the second quarter of 1983, government liability was estimated at around £7.5 million, but the estimates were approximate and unlikely to stabilise until all the aircraft had been sold. At that point, the government decided to resort to a loophole in order

41 Robert Ledger, "From Solidarity to 'Shock Therapy'. British Foreign Policy Towards Poland Under the Thatcher Government, 1980-1990", *Contemporary British History*, Vol. 30, No. 1 (2015), p. 99-118.

42 See John Campbell, *The Iron Lady*, London, Pimlico, 2004, p. 282-285, 339-343.

43 Margaret Thatcher, *The Downing Street Years*, London, Harper Collins, 1993, p. 790.

44 UKNA, FV 17/145: Wilde to Farrow, 03.04.79; UKNA, FV 17/341: note by Chilton, 02.05.79.

to get out of their financial responsibility. The underwriting agreement stated that Whitehall's assent was required before proceeding to build any aircraft after No. 10. However, the construction of No. 11 had begun in 1977 without BAe supplying cost estimates in advance and without any formal approval, as at that time officials thought this would facilitate the ROMBAC negotiations. Now, just as the Hurn line was finally closing, the DTI decided to use this technicality once an advantageous deal had been reached with BAe on other issues, although, months later, they had to admit that that line would not have been sustainable if the company had resorted to arbitration. Similar misunderstandings also occurred with regard to the suspension of the licence fee claim, which BAe believed the government had accepted in principle on conclusion of the ROMBAC negotiations. In short, relations remained tense and, in the summer of 1984, BAe's financial director, Bernard Friend, accused the government of trying to get out of its commitments by having recourse to loopholes that, for years, nobody had ever dreamt of using.[45]

There were also tensions in Romania. The ROMBAC programme included the transfer of basic specifications by BAe by the end of 1980 and the creation of a joint project management team to interface with IAvB (Întreprinderea de Avioane Bucureşti, which in 1978 had absorbed IRMA). IAvB would take care of both the Islander and the 1-11, while also supervising all the necessary procedures to start production. The rate of production was to be three planes a year starting from 1982, increasing to six in 1986, when the transfer of technology would be concluded with the completion of the 22nd unit. The materials would be supplied by BAe, which was to retain both design responsibility and product support in its entirety, as well as setting up a quality plan and guaranteeing its implementation. In this way, the CAA – whose inspectors would attend to monitoring – would be able to certify that aircraft produced in Bucharest were equivalent to those produced in the United Kingdom. A similar procedure would be applied between Turbomecanica and Rolls-Royce regarding the Spey. Of course, it was essential for a British authority to certify the Romanian aircraft and engines as fit for flight if the Romanians wanted to expand their export potential. This explained Bucharest's interest in the involvement of the CAA and the commitment to ensure

45 UKNA, FV 17/321: minute by Hudson, 25.07.83; UKNA, FV 17/322: Moorshead to Stanbrook, 22.12.83; Moorshead to Stanbrook, 20.01.84; UKNA, FV 17/323: minute by Rickford, 02.07.84; minute by Croft, 06.07.84. UKNA, FV 17/324: Arkell to Davies, 21.09.84.

that the procedures of the Romanian Civil Aviation Department should comply with British civil airworthiness requirements.[46]

Soon, however, the CAA inspector reported poor housekeeping standards at the huge factory that was being set up in the seven Băneasa hangars, along with other solvable problems. He also found that quality control was unsatisfactory, partly because the Romanians had tried to reproduce the sophisticated system in use at Weybridge, but without having sufficient qualified personnel, and partly because executives did not seem to be up to the task. Not only did the quality control officer and his deputy shut themselves away in their office, instead of interacting with the staff, but also the engineers involved in the inspections rarely showed up in the workshop. In addition, the Romanian executives belonging to the various organisations involved could hardly tolerate one another, while the quality surveyor sent by BAe in the spring of 1981 – James Ewen, a sixty-year-old engineer who had spent the 1970s in Toulouse, first on the Concorde and then on the Airbus A300 and A310 – was deemed ineffective. On top of it all, there were delays in communications between Bucharest and the United Kingdom due to the controls and censorship of the Romanian security services. A telex sent through IAvB took two or three days to get to the destination, so it is easy to understand why the most urgent or important communications were delivered through the embassy in the diplomatic bag.[47]

The CAA's concern about its position as a certification body, also considering the fact that BAe's commitment to the ROMBAC project had not come up to expectations,[48] finally turned into annoyance and embarrassment when, from 1981, the Romanians started shilly-shallying and then failed to pay what they owed. The DTI merely advised the CAA not to be too surprised given the party involved and not to make concessions:

> It would not be surprising to find in due course that some approach was made to us asking whether the hard currency element could not be offset or whether we could not forget about all the formalities and tell them that their work was acceptable. Should the latter happen, we would of course have to explain that business is not done in such a way in the United Kingdom.[49]

---

46 UKNA, DR 8/27: Terry to Chaplain, 16.08.79; UKNA, DR 8/28: note of a meeting held on 30.10.79; Răican to CAA, 14.02.80; Guilbert to Rain, 21.03.80.

47 UKNA, DR 8/28: comments on visit held on 13-17.10.80; UKNA, DR 8/29: minute by Inskip, 18.02.81; report on visit No. 2 held on 15-22.05.81; note by Inskip, 05.06.81; report on a visit held on 16-23.10.81.

48 UKNA, DR 8/29: meeting held at the CAA on 22.09.81.

49 UKNA, DR 8/29: Pownhall to Shelton, 15.09.81.

This was a pity because the CAA inspector had pointed out that, despite all the problems encountered, the assembly of the first four aircraft had got off to a good start, while BAe recognised the high standard of workmanship that could be achieved at Băneasa. However, in the summer of 1982, with Romania's failure to pay up, the CAA informed BAe that it intended to back out. The company's attitude in these circumstances is enlightening, as it noted that, without the CAA's involvement in certification, one of the prerequisites on which the licence agreement was based, disappeared. However, the company did not intend to take on the costs because their contractual obligation was to ensure that the aeroplane produced by the Romanians was equivalent to the British one and not to have the CAA certify it.[50]

At this time, rumours were flying around at BAe about abandoning the project, due to bad debts, the difficulties of achieving counter-trade and the accumulating delays. The last two aspects were incontrovertible. Bureaucracy and, often, the poor quality of Romanian goods had made it impossible to implement a strategy that was based on numerous small transactions on the British market and BAe had therefore sought to involve the Romanians in large contracts with third parties. For instance, in 1981, it tried to involve them – through the Lazard bank and a Lebanese agent – in the construction of a cement plant in Qatar, causing turmoil in Whitehall, where it was feared that the Qataris would interpret the initiative as a weakening of Thatcher's anti-Communism and would reconsider the profitable purchase of Rapier missiles from the BAe military division.[51] That being said, however, it should be noted that from January 1980 to February 1982 the Romanians had paid less than £14 million for the licence and that, at the end of 1981, BAe expected £4.6 million net profit from the ROMBAC project.[52]

BAe's and Whitehall's attitudes had changed latterly. Of the £190 million funded by the ECGD to secure the project, less than half of the money had been spent, but Romania had already amassed a backlog of £4 million in interests. Given that trade relations with countries behind the Iron Curtain were not promising and considering Romania's reorientation towards the CMEA, it was important to limit the United Kingdom's

[50] UKNA, DR 8/29: report on visit No. 2 held on 15-22.05.81; UKNA, DR 26/23: note on informal meeting to discuss ROMBAC 20 held on 08.07.82.

[51] UKNA, FCO 8/4116: Day to Palmer, 04.10.81; Willson to Palmer, 08.10.81; Day to Passmore, 12.10.81; minute by Miers, 19.10.81.

[52] UKNA, FV 17/319 Moorshead to Stanbrook, 14.08.82; minute by Stanbrook, 31.08.82; Arkell to Walters, 22.09.82.

exposure – which was already feeling the effect of the Thatcher therapy – to a partner who was in a precarious financial situation. It was therefore suggested that BAe and Rolls-Royce should agree with the Romanians to close the project, without – of course – this appearing to be the result of a British government initiative.[53]

The concrete prospects of actually terminating the ROMBAC project consensually were, however, non-existent. On the occasion of the visit to Bucharest of the Parliamentary Undersecretary of State for Foreign and Commonwealth Affairs, Malcolm Rifkind, in autumn 1982, the Romanians made it clear that, for them, the ROMBAC project continued to enjoy top priority, as demonstrated by Ceauşescu's presence during the inaugural flight of the first completed plane. Even if the project was closely linked to the dictator's prestige – according to an FCO expert – forthcoming changes in the regime were not to be expected, since in Romania there were none of the elements that had worked in Poland. Furthermore, the investments made were of such magnitude that even a different government would hardly have accepted liquidation.[54]

On the contrary, the Romanians used Rifkind's visit to emphasise their interest in the revival of the 1-11 by equipping it with a new engine, a prospect that BAe Weybridge Division was actually considering.[55] In fact, the Spey engine had a very loud thrust and in the 1970s an environmental protection regulation aimed at reducing noise pollution had been mooted – from Annex II to the ICAO Convention on International Aviation. Consequently, hush kits for the Spey had been developed, consisting of acoustically-treated duct linings and jet noise suppressor units, mounted as standard on the first Series 500 1-11 produced for TAROM in 1976.[56] There was, however, a more ambitious opportunity to develop a new engine that, taking advantage of the work on the RB.211, would have a larger bypass ratio and a lower core jet velocity, thus producing less noise. The United States was interested in this type of engine, which it intended to use for the Gulfstream IV, and Rolls-Royce developed it under the name of RB.183 Tay. With a new 44-inch fan, the bypass ratio went from 0.7 of the Spey to 3: if the Tay had been used on the 1-11, the plane would have met the more restrictive noise rules introduced by ICAO (Chapter 3 ICAO Annex 16) and by the American Federal Aviation Administration (Stage 3 FAR Part 36) just when the ROMBAC negotiations had begun,

53 UKNA, T 437/76: minute by Hawtin, 17.09.82; Harding to Hawtin, 01.10.82.

54 UKNA, T 437/76: Broomfield to Hawtin, 06.10.82; Broomfield to Adams, 03.11.82.

55 UKNA, T 437/76: Holmer to FCO, tel. 197, 13.10.82.

56 "Rolls-Royce Engine Update", *Aircraft Engineering*, Vol. 48, No. 1 (1976), p. 5-8.

prolonging its life until the end of the century and sharply increasing its sales prospects. In fact, with 15% less fuel consumption and an increase of 25% in range, the new 1-11 could also be used on night flights and on a wider variety of routes.[57]

In May 1983, the CNIAR general manager, Teodor Zamfirescu, asked the British to modify the licence with Rolls-Royce so that Turbomecanica could produce parts of the Tay, which would be mounted on the ROMBACs by IAvB.[58] To do this, Turbomecanica had to obtain the rights and know-how to produce a new nacelle. It seems that the Romanians had not considered that the Tay had not been developed from the Spey model with which they were more familiar, the 512DW, and that they had underestimated how little Rolls-Royce wanted their involvement. Past experiences did not indicate a willingness to relaunch relations with Romania and, furthermore, the priorities of the British government – still the owner of the company – were others. In fact, having been removed by the National Enterprise Board in 1980 and having reduced its net loss, Rolls-Royce's profit margins continued to be eroded by rising costs and debt, while concentration on large engines exposed it to the formidable American competition, especially after the Tristar was cancelled following Lockheed's exit from the civil aviation sector. For Thatcher's government, therefore, the only alternative to a slow decline softened by a public money outflow was to push Rolls-Royce to join one of the American giants (Pratt & Whitney, with the formation of International Aero Engine), so as to make it profitable again and privatise it.[59]

Prospects were no better on the BAe side. The company, privatised in February 1981, saw an increasingly important share of its profits from military systems, while in the civil aviation sector the focus was on the first airplane produced by BAe,[60] the 146 – the old HSA project cancelled in 1974 and that officials in Whitehall had scrapped as a possible alternative for collaboration with the Romanians in 1976 as it would have meant mounting American engines on it. A 1-11 propelled by the Tay would compete with the 146 in the same market segment and, therefore, with the exception of Weybridge, the company as a whole was very reluctant

57 UKNA, SUPP 29/297: "Further One-Eleven Development", n.d.; "Tay-powered BAC One-Eleven Proposed", *Aircraft Engineering*, Vol. 55, No. 8 (1983), p. 14-16.

58 Ken Goddard, *The Rolls-Royce Tay Engine and the BAC One-Eleven*, RRHT Historical Series No. 30, Derby, 2000, p. 19.

59 Lloyds, HO/GM/CRA.34: minute by Williamson, 04.03.82; UKNA, CAB 130/1186 MISC 25(82)3, Rolls-Royce Long Term Strategy, 05.03.82; MISC 25(82) 2nd meeting held on 25.05.82.

60 UKNA, FV 17/145: Dorman to Jardine, 20.10.76.

to implement this plan, sponsoring the 146 even in those cases when the ROMBAC seemed to stand a chance of being purchased.[61]

In addition, the interest shown by airlines did not seem to justify the risk. Noise regulations would force British Airways to renew its fleet (which included twenty-five 1-11s without hush kits), but the company was more inclined to choose the Boeing 737 as an intermediate solution, while there was strong political pressure on BA to buy A320s in the future, as proof of its 'Europeanism' seeing that it had never bought an Airbus aircraft. British Caledonian and Dan Air had said they were not interested in re-engining, preferring to resort to hush kits for some of their devices. US Air, which owned twenty-six 1-11s, was willing to consider purchasing new aircraft propelled by the Tay, but not to change the engines of the old ones.[62]

A re-engining programme would eventually be negotiated in 1985 by BAe with the American company Dee Howard – backed by the money and interest of Saudi magnate Salem bin Laden – with a view to reconfiguring the executive jet version of the 1-11. The accidental death of bin Laden in 1988, the acquisition of Dee Howard by Aeritalia and the technical difficulties encountered would delay the programme until the beginning of the 1990s.[63]

Nevertheless, considering the insistence of the Romanians – who actually stated publicly in September 1984 that by the end of the year they would get the licence to produce the Tay[64] – BAe, or at least that part of it linked historically to the 1-11, gave the impression of seeking the government's involvement in saving the project. Of course, this could be mere contractual scrupulousness, to ensure that the payments due, though somewhat intermittent, were still coming in, even in the darkest hour of Romanian austerity. However, when Charles Arkell, the commercial director of the Filton Division, spoke of a sort of moral responsibility towards the Romanian aeronautical industry, an alarm bell rang in Whitehall, where the Treasury, and in particular the ECGD, were

---

61 Lloyds, HO/GM/CRA.81: note on British Aerospace Plc., n.d.; UKNA, FV 17/321: Jenkins to Brooks, 08.09.83; Balmer to Brooks, 06.09.83; Burton to Browne, 05.10.83.

62 UKNA, FV 17/322: minute by Ralph, 06.04.84; UKNA, FV 17/334: meeting with the Chairmen of BAe, RR and BA held on 14.06.83; British Airways: A320, the Tay engine and the Noise Ban, 24.08.83.

63 UKNA, FV 17/322: note of a meeting held on 13.06.84; Goddard, *The Rolls-Royce Tay Engine*, *op. cit.*, p. 20-21, 31, 35-37.

64 "Farnborough", *InterAvia Newsletter*, No. 10,584, 10.09.84, p. 2.

determined not to increase exposure to Romania. On the contrary, with much reluctance, the funds already allocated to the Spey could be used on a new Tay-centred programme, but only to avoid the risk that, in the event of a complete breakdown, the Romanian side could suspend loan repayments as retaliation.[65]

In this rather chaotic situation, in the last months of 1986, Dee Howard had begun negotiating with the Romanians for their involvement in re-engining, but talks were disturbed by the activism of Derek Lowe, the director of Executive Jet Sales, an agency specialised in the marketing of the medium-sized corporate jet HS 125, which he founded when he started working as a freelancer in 1980.[66] Lowe was convinced that the 1-11 on which to mount the Tay should be the Series 500, which would have a full transatlantic capability. Unaware of the fact that the agreements between BAe and Dee Howard called for an exclusive with the latter for the possible re-engining of all the aircraft models, Lowe began a personal campaign in favour of a programme focused on safeguarding the British airframe industry and its components – one of the few sectors, in his words, where the fearsome Japanese competition was not strongly felt. Bombarding with letters for years the BAe executives and DTI officials, while also finding a possible source of funding in James Hanson, a businessman with a corporate rider fame, Lowe explained how the re-engining of Series 500 would save the ROMBAC project from bankruptcy and constitute the first step towards establishing an Anglo-Franco-Romanian trilateral cooperation to create a European core of experience in propfan engines. The British would contribute with a good fuselage like the one of the 1-1, Aérospatiale would not have to develop a plane on its own and both countries would benefit from the low price of Romanian metal work. Crucial competencies in the civil aviation sector would thus be maintained and, in the future, Airbus could use them to compete effectively with the Americans.[67]

For the DTI and BAe, Lowe's proposals were water off a duck's back, but they were used by the Romanians to prolong the negotiations, without making a secret of the fact that they would prefer a direct agreement with the British, from whom they expected more favourable conditions,

[65] UKNA, FV 17/322: Potter to Field, 22.06.84. Papers obtained by the UK FCO pursuant the Freedom of Information Act (hereafter FOI/FCO): McKearney to BAe Deputy Chief Executive, 03.07.84; Adams to Coombe, 20.08.85.

[66] "New Business-jet Agency", *Flight International*, 30.08.80, p. 768.

[67] UKNA, FV 17/334: Bowder to Gillibrand, 23.06.86; UKNA, SUPP 29/297: Lowe to Michell, 10.06.87; Lowe to Michell, 25.06.87; Lowe to Smith, 18.08.87.

perhaps on the basis of that 'moral responsibility' that had been evoked by Arkell. Thanks to the intervention of the ubiquitous Pop, by the end of 1986, the Minister of Mechanical Engineering, Marin Nedelcu, assured to British Ambassador Hugh Arbuthnott that his government did not intend to consider Lowe's proposals. However, doubts remained as regards Dee Howard, in particular on the extent of the Romanian industry's involvement and, consequently, on how much hard currency it would cash in. So, Arbuthnott advised Whitehall to do what was necessary to bring negotiations to a happy ending soon. Otherwise, the ROMBAC project, which was still being defined by a DTI official as a 'flagship project of UK/Romanian trade and most important as a symbol of cooperation'[68] on the occasion of a visit to the United Kingdom by Deputy Foreign Minister Aurel Duma, would slowly fizzle out. However, the answer that the ambassador received was far from encouraging: the British government could only reassure the Romanians that the deal with Dee Howard was the only viable route, although it was still unclear how the project would be funded.[69]

Towards the end of 1987, Ceausescu's foreign debt extinction policy, leaving the aviation industry without funds, was causing new problems. Firstly, to save money, the Romanians had given up the original clause of the licence agreement with Rolls-Royce, which included monitoring and surveillance on Turbomecanica's production line – a move similar to the one they had made when they stopped paying the CAA, pushing it to withdraw from the project. In this case, however, the lack of monitoring would mean that the five Spey engines already completed, as well as the subsequent twenty-two of which the Romanians already possessed the components ready for assembly, could not be mounted on the ROMBACs nor receive CAA certification, as Rolls-Royce's endorsement would be missing. Although Romanian executives, including Ceauşescu himself, were going at it with guns blazing, complaining about the unsuitability of the investment, the situation was embarrassing also for the British, since those uncertified engines – in the event of an accident – would still be Speys, made with components branded Rolls-Royce.[70]

Secondly, since the original programme assumed that the technology assimilation process would be over by 1987 with the last of the twenty-two ROMBACs to be built with materials supplied by the United Kingdom, the

68 FOI/FCO: note of a call by Mr. Duma on 28.11.86.

69 FOI/FCO: Arbuthnott to DTI, 17.12.86; Arbuthnott to DTI, 22.12.86; DTI to Bucharest, OTTER No. 1, 09.01.87.

70 FOI/FCO: minute by Allen, 09.11.87; Paget to Potter, 22.11.87.

technical assistance agreement with BAe and the one on financial coverage would expire at the end of that year. However, since only seven aeroplanes had been produced, it was necessary to extend the agreements, but the Romanians refused to renew the existing line of credit with Lloyds, claiming that this would be in contradiction with the objective of extinguishing foreign debt and, rather, offered to finance technical assistance entirely through counter-trade, instead of only 30% of it as initially established – an offer that obviously BAe considered a smokescreen. The negotiations on re-engining came to a complete halt. Dee Howard had striven to put together a consortium: the American Continental Grain Company – already active in counter-trading with Romania – would provide the money to cover the purchase of twenty ROMBACs and one hundred Tays by the end of 1991, which would be assembled thanks to the technology provided by Dee Howard and then used by Guinness Peat Aviation. However, the Romanians continued to be the major obstacle, lacking the money needed for the initial investment (about $50 million) and insisting, rather, on arguing that it was up to the British – as they had sold obsolete technology to Romania – to finance those 1-11 updates needed to meet the market. CNIAR's general manager, Ion Petroaica, identified this last point as a pre-condition for the renewal of the technical assistance agreement with BAe, which, until then, had been funded through a British bank.[71]

Against this backdrop, although BAe's executives stated that support for the continuation of the project was now widespread, provided that a plausible agreement could be reached at least on the issue of the financing of technical assistance, Whitehall officials had every right to believe that BAe wanted to put the British government in the position of not being able to refuse concrete help. This had already happened in more than one circumstance in the past, but things were different now, as an official of the DTI's Foreign Trade division noted:

> Personally I am not convinced that there is sufficient support from eg. Air Division, FCO etc. to justify going further on a limb than we have done so far in our general pronouncements [...]. Since no Government Department has any money to assist the project further, it has to be a commercial decision for the companies whether they pursue it or not. We can point out to them that to walk away from the deal at this stage might not do their international commercial image a lot of good.[72]

[71] FOI/FCO: HMA's call on the Minister for Machine Building Industry on 08.08.87; minute by Lance, 10.11.87; minute by George, 27.11.87; meeting on 30.11.87 to discuss the BAC 1-11 ROMBAC Project, December 1987.

[72] FOI/FCO: note by Potter, 09.11.87.

Moreover, at that time, the FCO was completing the policy review of bilateral relations with Romania. In the light of a systematic lack of respect for human rights – never included among the priorities of the Romanian regime –, of the uncooperative attitude in the CSCE negotiations and of an unwillingness to move in the direction of possible transformations, it was recommended not to make any concession to Bucharest.[73] In any case, it was believed that, if the ROMBAC project were terminated, the Romanians would not have effective weapons of commercial retaliation or, anyhow, that they would have been too inefficient to do much damage. On November 25, 1987, Whitehall officials concluded that:

> [...] there would be no disposition on our part to strive officiously to keep the project alive if British Aerospace saw nothing to be gained commercially from continuing the project. It was agreed that there was no possibility of steering British Aerospace into a course which they did not want to take.[74]

Finally, it was decided that the Minister of Trade Alan Clark and BAe would send a message to Deputy Prime Minister Andrei and his government colleagues in Foreign Affairs, Mechanical Engineering and Foreign Trade to clarify the terms of the situation and its gravity. The message bore no fruit and, at the beginning of 1988, the Romanians refused to renew the loan agreement covering BAe technical assistance.[75]

The following months saw no progress, with Bucharest continuing to insist on an impossible direct deal with BAe and Rolls-Royce, excluding Dee Howard, on the grounds that updating the plane – including re-engining – was a British responsibility. Speaking to Conservative MP Jack Aspinwall – who visited Romania in the summer of 1988 as a member of the Inter-Parliamentary Union and as MP for a constituency near Filton – Ceauşescu said that the troubles afflicting the ROMBAC programme were attributable to BAe's unwillingness to accept Romanian goods for counter-trade, once again alluding to the possibility of direct intent. BAe, who had not recalled their technicians from Bucharest, despite the lack of adequate financial and contractual coverage, was ready to accept limited involvement in the renewed package offered by Dee Howard. A new partner, Contitrade Services (Belgium), would fund the project, starting with the purchase of a ROMBAC that would be used by Dee Howard as a re-engining test bed and then returned to the Romanians with the Tays.

---

73 FOI/FCO: minute by Allen, 07.12.87.

74 FOI/FCO: minute by Lance, 25.11.87.

75 FOI/FCO: Clark to Andrei, 11.12.87; George to Arbuthnott, OTTER No. 24, 14.12.87; ECGD: Rombac position paper, n.d. [Spring 1988].

However, all the usual obstacles remained: first, the Romanians continued to oppose the principle of third party financing and, consequently, did not want Contitrade as a partner. Secondly, Dee Howard did not intend to produce the nacelles and the thrust reversers in Bucharest. Finally, while the Romanians did not want commitments after 1990 – when, in theory, the last of the twenty-two planes originally planned would be completed – BAe wanted a longer-lasting agreement to guarantee Romania's commitment to the re-engining programme.[76]

The pessimism of British officials about the outcome of these negotiations stood out a mile. David Potter of the DTI Foreign Trade Division pointed out that the ROMBAC project had been launched ten years earlier in a completely different political climate and was now virtually a no-hoper.[77] It seems surprising that, at this point, no one actually decided to pull the plug. Indeed, while BAe could rely on the internal lobbying activity of the Civil Aircraft Division, the British government could not fail to recognise that the Romanian foreign debt extinction policy was, at the time, beneficial to the United Kingdom. As the ECGD pointed out to Potter:

> Despite the lack of success of the project the Romanians are continuing to repay, and indeed are gradually prepaying their loans (£46.8 m of prepayment were received at the end of September alone). [...] Repayment and prepayment have now reduced the principal outstanding to £34.9 m (interests would be extra) or roughly 20% of the total loans of £172.7 m. Hopefully we shall see the balance prepaid in the next year or so, if not sooner, and the longer we can play for time the less that will be at risk to any financial retaliation. [...] Therefore at this stage we would propose that BAe should be reminded that our approval would need to be sought before any decision to terminate their contracts was taken.[78]

So, although the Romanians were supposed to be too disorganised to implement reprisals, the risk was not worth taking. Evidently, the need to recover state loans was more important than the principle of government non-interference in the economic decisions of industrial companies.

In fact, by the end of 1988 there seemed to be some progress, with the Romanians finally accepting the principle of third party financing through Contitrade as a precondition for a secure payment mechanism. However,

---

76 FOI/FCO: Arbuthnott to PS/Minister for Trade, OTTER 15, 19.04.88; Bucharest to DTI, OTTER 20, 11.07.88; Paget to George, 01.09.88.

77 FOI/FCO: minute by Potter, 22.07.88.

78 FOI/FCO: Rosenthal to Potter, 02.11.88.

several serious ambiguities remained. Ceaușescu, speaking with the Deputy Chief Executive Officer of BAe Civil Aircraft Division, Robert McKinlay, outlined a production programme for 1989-1991 that differed from that proposed by Romanian officials, arguing that the transition to the Tay would have to come about without rushing things. The immediate goal was to secure BAe's help to complete the original lot of twenty-two planes on which the Speys would be mounted. The Tays would only be used at a later date and, once again, Romania would not invest any fresh money. In practice, therefore, Ceaușescu wanted to force out Dee Howard and obtain, with some concessions, the support of BAe in persuading Rolls-Royce to accept the certification of engines produced by Turbomecanica. In March 1989, shortly before the announcement of Romania's foreign debt extinction, Rolls-Royce's financial director was assured that £600,000 arrears accumulated on the Viper licence would soon be paid off.[79]

Although it would be interesting to understand which market outlets the ROMBACs with the Spey could have, considering that for years the only agreement had been that re-engining was necessary to prolong the life of the aircraft, in the final months of Ceaușescu's regime, Romania seemed to be operating on two negotiating fronts simultaneously. In 1988, a British leasing company, Swift Aviation, had developed a plan for the purchase of up to fifty ROMBACs. Swift would create a consortium to buy green airframes produced by IAvB with materials and components supplied by BAe. The mounting of the Tays, the new nacelles and thrust reversers would be done in Bucharest, while fitting-out would take place in the United Kingdom, where Swift would arrange leasing. Although the advantage was to minimise the investment required from Romania, at least initially, the plan met with hostility and scepticism on all sides, both because of doubts about Swift Aviation's ability to handle such a programme and because of the reluctance of BAe and Rolls-Royce to embark on this new venture. Nevertheless, the project however took off in 1989, after further restructuring at BAe meant the cancellation of the Civil Aircraft Division in favour of a Commercial Aircraft subsidiary organised in three autonomous accounting divisions, all called upon to cut costs and ensure profit. With this new configuration, while, on the one hand, the residual interest in the 1-11 was further downsized through amalgamation with the Airbus Division, on the other hand, pressures increased for a short-term return. For its part, Swift roped in heavyweights,

[79] FOI/FCO: Bucharest to DTI, OTTER 27, 07.11.88; Paget to DTI, 08.11.88; Bucharest to DTI, OTTER 29, 11.11.88; meeting at the DTI with the Romanian Ambassador, 21.01.89; note by Potter, March 1989.

such as the former Minister of State for Industry, Geoffrey Pattie, and managed to overcome BAe's reluctance towards its project. BAe would thus deal with production management in Romania and would provide the know-how for re-engining by purchasing it from Dee Howard. After a first Memorandum of Understanding, which included the completion of the first Tay ROMBAC in August 1991, in October 1989 Romanian Foreign Minister Ion Ungur spoke of Swift – which had created the Associated Aerospace (AA) consortium – as a solid opportunity to relaunch the ROMBAC project.[80] The violent end of Ceauşescu's regime would, however, open up completely unforeseen scenarios.

6.2 Post-Revolutionary Demise

As foreseen a few years earlier, the ROMBAC project survived Ceauşescu – at least for a while. The revolution of December 1989 is an event rarely addressed in international historiography, despite its obvious centrality for the transition to a democratic regime in Romania and its political value as the legitimising myth for the National Salvation Front (NSF), that is, the grouping of former communists and dissidents who, with the support of the military, took power in the chaos following Ceauşescu's escape from Bucharest on December 22.[81] This continued relevance is also apparent from the use of terms with very different denotations to indicate the same events,[82] the work of two official inquiry committees in the first half of the 1990s[83] and the creation of the Institute of the Romanian revolution of 1989,[84] which preserves archival funds and has published significant contributions.[85]

---

80 FOI/FCO: Atkinson to DTI, 12.10.89. Goddard, *The Rolls-Royce Tay Engine*, *op. cit.*, p. 33-35, 38-42, 45; "BAe Restructures", *Flight International*, 07.01.89, p. 5.

81 See Ioan Scurtu, *Revoluția Română din decembrie 1989 în context internațional*, Bucharest, Ed. Redacției Publicațiilor pentru Străinătate, 2009.

82 Peter Siani-Davis, "The Revolution After the Revolution", in Duncan Light and David Phinnemore (eds.), *Post-communist Romania. Coming to Terms with Transition*, Basingstoke, Palgrave, 2001, p. 15-34.

83 See Sergiu Nicolaescu, *Cartea Revoluției Române din Decembrie 1989*, Bucharest, Ed. Ion Cristoiu, 1999 and Şerban Săndulescu, *Lovitura de stat a confiscat revoluția română*, Bucharest, Omega Press, 1996.

84 See Constantin Sava and Constantin Monac (eds.), *Revoluția Română din decembrie 1989 retrăită prin documente si mărturii*, Bucharest, Axioma, 2001 and Dumitru Preda and Mihai Retegan (eds.), *1989 Principiul dominoului. Prăbuşirea regimurilor comuniste europene. Documente*, Bucharest, Fundația Culturală Română, 2000.

85 See Ştefan Păun, Corneliu Mihai Lungu and Alexandu Oşca, *Revoluția Română din 1989 la Bucureşti*, Bucharest, Ed. IRRD, 2015.

What matters here is that the lack of an organised civil society, the strong presence in the NSF of members of the Communist nomenclature who had latterly fallen from grace with the dictator and, more generally, the dim view of their work during the transition, were all things that, in the eyes of British diplomacy, were pretty much inevitable and predictable,[86] without in any way denying the revolutionary nature of the events. For example, Ambassador Michael Atkinson wrote to Whitehall in his 1990 overview:

> I do not believe it is right to dismiss [the Romanian revolution], as The Times does, as 'a *coup d'état* designed to keep a reformed and renamed communist clique in power.' I sensed impending revolt in the autumn of 1989 and wrote in my First Impressions Despatch of 25 October that it could be hoped that the Ceauşescu regime would end 'in months rather than years.' I did not know that extra impetus would be given by events elsewhere and that Ceauşescu had only two months to live. The National Salvation Front [...] was dominated by dissident apparatchiks with military support. It also co-opted many dissident intellectuals most of whom soon drifted away disillusioned with the Front's perceived unwillingness or real inability to achieve an almost instant transformation of Romania's society, economy and administration. Such a shift after many decades of dictatorship and little earlier experience of democracy would have required a miracle. Many Romanian intellectuals in the heady circumstances of 1990 and with their national propensity to exaggerate entertained unrealistic high hopes.[87]

During this period, Romanian foreign policy dithered between the temptations of possible opportunities for cooperation in the Black Sea area and the Balkans and the attractions of the West with its prospects of integration into the European Community and NATO.[88] This was compounded by the volatility of the new regime with regard to the

---

86 Stephen Kotkin, *Uncivil Society. 1989 and the Implosion of the Communist Establishment*, New York, The Modern Library, 2009, p. 69-96; Raluca Grosescu, *Les communistes dans l'après-communisme. Trajectoires de conversion politique de la nomenklatura roumaine après 1989*, Paris, Houdiard, 2012; Tom Gallagher, *Modern Romania. The End of Communism, the Failure of Democratic Reform, and the Theft of a Nation*, New York, New York University Press, 2005, p. 70-119; Anneli Gabanyi, "The New Business Elite: From Nomenklatura to Oligarchy", in Henry Carey (ed.), *Romania Since 1989: Politics, Economics, and Society*, Lanham, Lexington Books, 2004, p. 353-372.

87 FOI/FCO: Atkinson to Hurd, 14.01.91.

88 See Ruxandra Ivan, *La politique étrangère roumaine (1990-2006)*, Brussels, Éditions de l'Université de Bruxelles, 2009, p. 112-123; David Phinnemore, "Romania and Euro-Atlantic Integration Since 1989: A Decade of Frustration?", in Light – Phinnemore (eds.), *Post-communist Romania, op. cit.*, p. 245-269; Valentin Stan, *România şi eşecul campaniei Vest*, Ed. Universităţii din Bucareşti, 1999; Ronald Linden, "After the

evolution of internal politics and economic reforms. The insistence on competence – in contrast with the capriciousness of recent years under Communist dictatorship – when it came to the assignment of posts, had the unfortunate effect of obscuring previous political affiliations.[89] In addition, the gradualism of the process of dismantling the planned economy certainly did not enhance Romania's image, which was already seriously besmirched abroad, especially in Britain.[90] The fact that the miners repeatedly resorted to street violence as a means of exerting political pressure might have been justified in private by the state's top brass, but it reduced the country's chances of modifying the disinclination shown towards Romania by many Western governments, even after Ceauşescu's downfall.

> I was told vary frankly that the [Romanian] government had not felt so able to do so after the revolution to rely on the police or the army. They had no choice but to appeal to their supporters who had voted for them in office, for support. There was little or no pretence that the support had been wholly spontaneous.[91]

At the time, the winning model for effecting the transition could have been a sort of corrective shock therapy. This would have made it possible for Romania to benefit from the growing availability of European markets and to catch up with the West more quickly,[92] but Romania took a different line. While it is true that some of the schematisms of the 1990s should be reconsidered[93] and that, after the sufferings experienced by the population during the so-called 'Golden Age,' it would have been very difficult for a heterogeneous political system without broad consensus to demand fresh sacrifices in the name of possible future benefits. However, the gradualism characterising the years of Ion Iliescu's presidency was at least partly determined by the requirements of the industrial establishment and it often allowed undeclared private interests to take hold, though

---

Revolution: A Foreign Policy Bound to Change", in Daniel Nelson (ed.), *Romania after Tyranny*, Boulder, Westview Press, 1992, p. 203-238.

89 Peter Siani-Davies, *The Romanian Revolution of December 1989*, Ithaca, Cornell University Press, 2005, p. 215-219.

90 *Revoluţia Română văzută de ziarişti americani şi englezi*, Bucharest, Editions Evenimentul, 1991.

91 AMEJ-1-10-47 part 2: Amery to Hurd, 24.08.90.

92 Wendy Carlin and Michael Landesmann, "From Theory into Practice? Restructuring and Dynamism in Transition Economy", *Oxford Review of Economic Policy*, Vol. 13, No. 2 (1997), p. 77-105.

93 See Eichengreen, *The European Economy Since 1945*, *op. cit.*, p. 304-314.

without benefiting the transition as a whole. With a decline of 54% in industrial production and 25% of GDP between 1990 and 1992, Romania had the worst performance of the six European countries of the CMEA, while the reversal of the policy followed by Ceauşescu – for political and consensus reasons – led to a rise in domestic consumption, all at the expense of the formation of gross fixed capital and current account balances. This trend depleted the foreign exchange reserves that had been laboriously accumulated in the 1980s – which might have enhanced Romania's creditworthiness – to the point where, by the end of the decade, the country was again on the brink of default. Conversely, the reluctance to implement reforms prevented the country from dealing with the sad economic legacy of Communism: high capital stock obsolescence and low proportion of human capital intensive exports compared to other countries in the bloc. While the technological structure of exports to the EC countries had already highlighted the failure of Ceauşescu's forced economic development in 1988, the gap with states such as Poland and Hungary widened even further in the first half of the 1990s.[94]

The file on the restructuring and privatisation of state-owned enterprises is particularly enlightening. Law No. 15 of August 7, 1990 – which came into effect the following month, almost a year after the revolution – provided for their transformation into joint stock companies or, in the case of utilities or military production, into non-privatisable public companies that enjoyed almost total protection from creditors and were not subject to stringent budgetary constraints.[95] The former should have been privatised, but it took another year before Law No. 58 of August 14, 1991 laid down the procedures to be followed.[96] Not only were the criteria distinguishing between the two business groups unclear, but the privatisations continued to suffer constant setbacks, while the reluctance

---

94 David Turnock, *Aspects of Independent Romania's Economic History with Particular Reference to Transition for EU Accession*, Cheltenham, Ashgate, 2013, p. 93-99; *Id.*, *Restructuring the Romanian Economy: A Study of Transition 1990-2006*, Elgar, 2007; Alan Smith, "The Transition to a Market Economy in Romania and the Competitiveness of Exports", in Light – Phinnemore (eds.), *Post-communist Romania*, *op. cit.*, p. 127-150; Teodor Nicolaescu, "Privatization in Romania. The Case for Financial Institutions", in Donald Fair and Robert Raymond (eds.), *The New Europe: Evolving Economic and Financial Systems in East and West*, Dordrecht, Springer, 1993, p. 101-109; Catherine Durandin and Zoe Petre, *Romania Since 1989*, New York, Columbia University Press, 2010, p. 172-173.

95 Legea No. 15/1990 privind reorganizarea unităţilor economice de stat ca regii autonome şi societăţi comerciale, în vigoare de la 07.09.90, available at http://lege5.ro/.

96 Legea No. 58/1991 privind privatizarea societăţilor comerciale şi vânzarea de active din patrimoniul acestora, în vigoare de la 16.08.91, available at http://lege5.ro/.

to sell control packages to foreign investors betrayed an underestimation of the role of foreign direct investment in triggering the process of economic transition.[97]

In particular, the fact that Western countries were not prepared to risk large amounts of capital in an international recession was apparently not taken into account. There was no chance of repeating what had happened in the 1970s, when Western credit availability had fuelled the increase in interchange and, at least in some countries, the consumption of goods: the bulk of the money had to come from the internal restructuring of transition economies and those countries that were quicker to put this into practice attracted more investment.[98] This, moreover, was consistent with the economic paradigm that had emerged in the 1980s, well-documented by the activities of the European Bank for Reconstruction and Development (EBRD), which was founded in 1991 for the purpose of contributing to the development of the countries of Central and Eastern Europe. Apart from the obstacle of internal conflicts, the bank refused to lend money to the public sector, seeing it as a kind of subsidy to what remained of the economic failures of Communism.[99] So, although post-Ceauşescu Romania needed help, its behaviour only exacerbated the West's distrust of the new leadership and, certainly, Romania's actions did not make it any easier for the country to get their hands into their pockets.[100]

The same was true for the United Kingdom, whose relations with Romania had run out of steam in the last years of the dictatorship, with no reversal of the trend after the revolution. In the list of international agreements signed by Bucharest between 1990 and 1994, the United Kingdom is conspicuous by its absence, nor was there

97 See Turnock, *Aspects of Independent Romania's Economic History*, *op. cit.*, p. 94-97; Dragoş Negrescu, *A Decade of Privatization in Romania*, Bucharest, Romanian Centre for Economic Policy, 1999; Raphael Shen, *The Restructuring of Romania's Economy. A Paradigm of Flexibility*, London, Praeger, 1997, p. 121-158; Avner Ben-Ner and John Montias, "Economic Systems Reforms and Privatization in Romania", in Saul Estrin (ed.), *Privatization in Central and Eastern Europe*, London, Longman, 1994, p. 279-310.

98 Steven Morewood, "The Demise of Command Economies in the Soviet Union and its Outer Empire", in Michael Oliver and Derek Aldcroft (eds.), *Economic Disasters of the Twentieth Century*, Chatenham, Elgar, 2007, p. 302-304.

99 Adam Bronstone, *The European Bank for Reconstruction and Development. The Building of a Bank for East-Central Europe*, Manchester, Manchester University Press, 1999, p. 26-92; Steven Weber, "Origins of the European Bank for Reconstruction and Development", *International Organization*, Vol. 48, No. 1 (1994), p. 1-38.

100 Bache, BDOHP, available at www.chu.cam.ac.uk (last accessed on July 29, 2015).

any Intergovernmental Cooperation Committee,[101] while the Know-How Fund, considered to be the jewel in the crown of a 'diplomacy of expertise,' whose aim was to secure the low-cost transition of the countries of Central and Eastern Europe, was extended to Romania only in 1991, after overcoming serious doubts concerning the prospects of democracy and the free market in the country, and focusing on training in banking, agriculture and diplomacy.[102]

Given the circumstances, the fate of the Romanian aeronautical industry and, in particular, the ROMBAC project was anybody's guess. After the openings and co-operatives of the 1970s, ten years had been lost and only nine aircraft had been completed. Indeed, the most concrete perspective was the Associated Aerospace consortium organised by Swift, to which BAe had given its support just three weeks before Ceauşescu's execution.[103] The problem, of course, was how to put together the money and the consensus needed to get the project off the ground. The indomitable Pop went to Romania for three weeks shortly after the revolution and secured a private invitation for the former Conservative Minister of Aviation, Julian Amery, to visit the country. The invitation came from former Romanian Foreign Minister Gliga – whose letter describing the uncertainty of the situation and the lack of inclination towards full liberalisation of the economy was delivered to Amery by Pop himself – and his successor-to-be, Sergiu Celac (the latter, not having a formal position, invited Amery in his role as President of the Romanian Sporting Hunters and Anglers Association).[104] The interpreter that Ceauşescu used during his 1978 state visit had had occasion to know and appreciate Pop during the preparations for Bodnăraş's trip to the United Kingdom in 1973.[105]

Having consulted the FCO, Amery agreed to go to Romania and meet the most important members of the 'Communist mafia',[106] namely the NSF, between February and March 1990. He found serious political instability and even did not discount the possibility of a military coup in the event of a major uprising during the elections scheduled for May 1990.

---

101 See Adrian Năstase, *Acordurile internaţionale încheitate de România sau la care România a devenit parte (1990-1994)*, Bucharest, Metropol, 1995, p. 144-147.

102 FOI/FCO: Atkinson to Hurd, 15.01.92. See Keith Hamilton, *Transformational Diplomacy after the Cold War. Britain's Know-How Fund in Post-Communist Europe, 1989-2003*, Abingdon, Routledge, 2013, p. 49-52.

103 FOI/FCO: note by [Potter] OT3/5B, 13.12.89.

104 AMEJ-1-10-47 part 3: Gliga to Amery, 21.01.90; Celac to Amery, 24.01.90.

105 Celac to the Author, 29.03.15.

106 AMEJ-1-10-47 part 3: Amery to Deakin, 29.01.90.

He also noted that British diplomatic representation was at an all-time low, unable to keep up with the activism of the French and Germans.[107] Above all, Amery took it upon himself to act as a mediator for Rolls-Royce. He wrote to the president of the company, underlining the Romanian balance of payments surplus as a positive credential for potential investors, and, last but not least, the usefulness of Pop as an intermediary in the event of the probable failure to replace the ruling class.

> It will take time for a political class to develop in Romania. Meanwhile much the same people will remain in positions of power and influence, as in the last regime. George knows them all and since Romania has about the best credit rating of any of the emerging democracies your prospects of doing business should be better than most. The old vampire [Ceauşescu] had an obsession about paying his debts. His people paid a high price for it but it is a valuable legacy for them.[108]

However, Francis Tombs' reply was far from encouraging:

> Their aerospace industry represents an asset on which they must build, but we have stressed the importance of being realistic in their ambitions and of honesty about quality if they are going to be accepted as a supplier. We will do what we can to help them help themselves. What we are not prepared to do is enter into a Spey type licence deal for the Tay engines they will need if their ROMBAC 1-11 programme is to succeed; these they must buy and hopefully their credit rating will help them with that programme.[109]

The attitude of the British government was also substantially negative and at the FCO it was clear that Romania could soon be on the brink of anarchy. It was therefore out of the question to think of using public money to assist the relaunch of the ROMBAC project, as requested by the BAe Airbus Division to FCO Ministry of State William Waldegrave. Rather, as Secretary of State Douglas Hurd told Amery, the long-term perspective of economic aid ought to serve as an incentive to push the Romanians to implement reforms in favour of democracy and the free market. That would be a bit like dangling the carrot in front of the peacock's eyes to make it walk; first, there had to be reforms and new sacrifices, and only later, if at all, the money. In fact, the ECGD excluded a recovery in the medium-term coverage of transactions with Romania before the spring

---

107 AMEJ-1-10-47 part 2: Amery to Hurd, 26.03.90.

108 AMEJ-1-10-47 part 3: Amery to Tombs, 16.03.90.

109 AMEJ-1-10-47 part 3: Tombs to Amery, 29.03.90.

of 1991, therefore, only after the IMF's exploratory mission in autumn 1990 would it provide any reliable data on risk.[110]

It is therefore hardly surprising that the official launch of the Associated Aerospace plan on June 30, 1990 was met with scepticism: the contract specified many conditions for its enforcement. Rolls-Royce had requested payment of an early deposit before signing the agreement to supply the Tay engines; also, production targets were judged unrealistic by British officials and, above all, the financial aspect remained rather vague. Who would put up the money? The fact that the bank involved was Duménil Leblé, which had been in serious difficulty since getting caught up in the failed attempt by the entrepreneur Carlo De Benedetti to try his luck with Société Générale de Belgique, did not augur well.[111]

The Romanians themselves were not fully convinced of its success; almost at the same time as the signing of the contract, they were considering possible alternative solutions, such as an agreement with British company A.J. Walter, whereby the latter would become a non-exclusive agent for ROMBAC sales and rentals. The planes would be powered by Spey engines equipped with a special hush kit to be developed by Quiet Nacelle Corp., Miami, Florida. Rolls-Royce, who had been asked for advice on this issue, pointed out to the Ministry of Mechanical Engineering that it would be impossible to bring the Spey to the same noise levels of the Tay, while the loss of performance and hush kit development costs would make the project uneconomical.[112] In any case, according to Eugeniu Smirnov, managing director of IAvB, the price set by Associated Aerospace for the plane would be unprofitable for the Romanian company. Unfruitful negotiations ensued on the financial terms that would be applied; at the same time, the Romanians appointed the financial consultants Coopers & Lybrand to assess the soundness of Associated Aerospace. The resulting report, which was highly critical, gave the Romanians further pause for thought, until in March 1991 a British court ordered the liquidation of Associated Aerospace for unpaid debts.[113]

---

110 FOI/FCO: note by OT3/5B, 04.01.90; Dep. Managing Director (Operations) BAe Airbus Div. to Waldegrave, 16.02.90; ECGD to the 1st Secretary (Commercial) Bucharest, 19.09.90; AMEJ-1-10-47 part 2: Hurd to Amery, 07.09.90.

111 FOI/FCO: Atkinson to DTI, OTTER 32, 09.07.90; 1st Secretary (Commercial) Bucharest to OT3/5B, 17.08.90.

112 Papers obtained from the UK Department for Business, Innovation & Skills pursuant the Freedom of Information Act (hereafter FOI/DTI): Rolls-Royce Director of Commercial Services to the Ministry of Machine Building Industry, 13.06.90; "Romania Signs New One-Eleven Deal", *Flight International*, 17.07.90, p. 5.

113 FOI/FCO: Atkinson to DTI, OTTER 52, 11.10.90; note by OT3/5B, 05.03.91; Goddard, *The Rolls-Royce Tay Engine*, *op. cit.*, p. 69-77.

When interviewed by Potter, BAe Commercial Aircraft's managing director, Maurice Dixson, glossed over the Memorandum of Understanding between BAe and AA in 1989, stating that the latter had never been contractually or financially linked to his company. The core of the issue, according to Dixson, was that the Romanians had to face up to the financial burden of restructuring their aerospace industry and that they should not expect others to pay for them.

> BAe remain 'sympathetic' to the Romanians as licensees of the 1-11. However BAe do not see it as part of their role to effectively restructure the Romanian aircraft industry and refinance the ROMBAC project. BAe acknowledge they have a general 'moral' responsibility to the Romanians as licensees of the 1-11; BAe would not walk away from this. [...] Unfortunately the Romanians had delayed too long with ROMBAC and missed the boat or plane. The latest development in the civil aviation sector mean that the 1-11 could prove difficult to sell or lease: however good, it is still an older generation aircraft. The Romanian are aware of this but reluctant to admit it.[114]

From the information available, it is not possible to state with certainty how far the Romanians really expected more concrete help once the dictatorship had fallen. Certainly, however, both the leaders of Romaero, as the IAvB had been renamed, and the head of the Department of Mechanical Engineering of the Ministry of Resources and Industry, Andrei Pintilie, began to see a more realistic future for subcontracting and repairs to major Western companies, perhaps through joint ventures that would still leave Romania's government in control of Romaero. With this in mind, President Ion Iliescu had personally had frequent meetings with Boeing representatives during 1990 in order to assess how Boeing might use the Romanian structures to produce some of its components.[115]

The British Government made BAe's position clear to the Romanians in a letter from the Minister for Corporate Affairs, John Redwood, but at the same time, Bucharest asked BAe to prepare a new business plan. Finalised in August 1991, this focused on the construction of the last twelve uncompleted aircraft of the initial series of twenty-two, using the materials already available in the Romaero warehouses and exploiting the re-engining work done by Dee Howard. As the plan was based on the

---

114 FOI/FCO: minute by OT3, 18.03.91.

115 Papers obtained from the U.S. Department of State pursuant the Freedom of Information Act (hereafter FOIA/DoS): Bucharest to SecState, No. 10418, 05.10.90; FOI/FCO: Atkinson to DTI, OTTER 50, 05.10.90; "Romania Looks for Western Investment", *Flight International*, 10.07.90, p. 13; Alan Postlethwaite, "Coming in from the Cold", *Flight International*, 17.07.90, p. 26-28.

principle of division of business risks, the Romanian government was asked to create adequate credit facilities and to reach an agreement with Alenia, now controlling Dee Howard, on the financial arrangements needed to gain access to know-how. For its part, BAe requested the coverage of the ECGD, already demanded by Rolls-Royce for the sale of the Tay to the Romanians.[116]

Ambassador Atkinson, after contacting BAe in Bucharest, actually worked hard with both the DTI and directly with the ECGD, pointing out that this much smaller project had good chances of success. A government contribution in the form of a guarantee for exports to Romania, not necessarily for the full £100 million, would give the Romanian industry an opportunity for growth and would increase BAe's chances of obtaining the modernisation work for Otopeni Airport.[117]

It would probably have been very difficult to obtain ECGD coverage, but the project actually foundered on the industrial side as Alenia demanded $50 million to run the Series 500 Tay ROMBAC certification programme in Italy, in order to recover the money lost through the delays and bad prospects of Dee Howard's programme. By the end of the year, the latter, which was crucial to the relaunch of the ROMBAC line, was definitively cancelled, while Alenia filed a lawsuit against BAe on charges of negligence. Furthermore, the slow pace of the 1990-1991 flight tests on Dee Howard's prototype seemed like a portent of doom.[118]

The collapse of Dee Howard and the Associated Aerospace projects was apparently the finishing stroke for the ROMBAC. In 1992, BAe had seemed ready to pull the plug once and for all, recalling from Bucharest their staff still in Romania, but a glimmer of hope remained: if the Romanian government had given the Băneasa company the necessary financial support to guarantee covering the costs, BAe would have continued to support the project.[119]

In this climate, like a bolt from the blue, the Romanian media and, then, the international press published some sensational news: on February 7, 1993, a small American company in Newark, Kiwi Airlines,

---

[116] FOI/FCO: Brooks to Atkinson, 13.08.91.

[117] FOI/FCO: Atkinson to the Director of the DTI Soviet and East European Branch, 09.08.91; Atkinson to ECGD, 16.08.91.

[118] U.K. Civil Aviation Administration, Gatwick, file No. 9/33/7037/A18404: BAC 1-11 Tay Power Series 2400 – meeting notes, 21.03.90 – 13.11.91; Goddard, *The Rolls-Royce Tay Engine*, *op. cit.*, p. 84-86, 93-94.

[119] FOI/FCO: note by the DTI European Division, 03.04.92; FOI/DTI: Bucharest to FCO, tel. 568, 17.12.92.

operating since September 1991 on some domestic routes with a couple of rented 727s, signed a frame contract with Romaero for the purchase of eleven Tay ROMBACs – again to complete the original lot of twenty-two planes – plus an option for five other aircraft. It appeared that no one had been informed of the negotiation, which lasted only a month, and the details reached the British government mainly through Romaero, either by speaking directly to the Sales and Marketing Manager Constantin Dinischiotu, or indirectly through the two BAe and Rolls-Royce contacts, Peter Collins and Ken Goddard, respectively the ROMBAC Project Manager and Project Marketing Manager for Small Engines. BAe would once again retain design authority, but would avoid any embarrassing incompatibilities with its partnership with Taiwan Aerospace for the production of the 146, leaving the role of leading engineering company to another British company. Rolls-Royce, after some initial scepticism, began to show interest in a relaunch that was technically feasible and that, if it were extended to re-engining TAROM's 1-11s, would allow the company to sell seventy Tays at $2.3 million each.[120]

This was good news, coming at a time when the Romanian government appeared to be willing to consider a serious revival of the aviation industry, perhaps in view of a resumption of privatisation.[121] The problem, again, was to put together the necessary money. Through the services of American aviation financial consultant Philip Bloom, a $15 million bridging loan was prepared by Bankers Trust, a company specialising in capital markets, in particular in derivatives, but funding was required to cover the entire amount needed for the project (about $100 million over five years). Lloyds was working on this, apparently with good prospects, but they thought that a list of lenders could only be drawn up if ECGD coverage could be counted on. This meant that the ECGD would have to change both the maximum ceiling for Romania (£30 million) and the one for each project (up to £10 million). Rolls-Royce tried to convince the ECGD, explaining that the Kiwi project was more realistic than its previous versions and that, if it worked, it might stimulate a substantial international demand for

---

120 FOI/FCO: note by the DTI European Division, 24.02.93; Bucharest to DTI, INTRA 4, 17.02.93; FOI/DTI: note by the first secretary (commercial) Bucharest, 08.02.93; record of a visit to H.M. Embassy by RR representatives on 18.03.93; FOIA/DoS: Bucharest to SecState, No. 1155, 08.02.93; "Romanians in Deal To Build 11 BAC 1-11s", *Financial Times*, 09.02.93; "FLS Favourite in One-Eleven Tay Project", *Flight International*, 25.05.93, p. 8.

121 FOI/FCO: note by the DTI European Division, 15.02.93; FOI/DTI: FOI/FCO: note by the DTI European Division, 23.02.93; Kate Sarsfield, "Romanian Reality", *Flight International*, 27.07.93, p. 33.

the Tay ROMBAC. However, the ECGD's response was negative; they maintained that even if suspending the plafond for a single project was not inconceivable, there was no realistic possibility of raising the overall ceiling for Romania.[122]

On July 7, there was a new turn of events: without the embassy's prior knowledge, BAe recalled its engineers from Bucharest, thus ending the license for the construction of the 1-11. It is very likely that this initiative, justified by the frustration of the company due to Romaero's insolvent debts and for the umpteenth relaunched that was slow to get off the ground, was actually motivated by the desire to put the Romanian government's back against the wall, in a way, forcing it to give a full financial backing to the project. Such a guarantee had actually been specified as a precondition from the outset – given the impossibility of both BAe and Rolls-Royce to expose themselves to a spiral of risky investments – and it was increasingly becoming a prime requirement to unblock the situation with the ECGD, in line with the principle that the money would only be forthcoming once the Romanians had contributed their share.[123]

Indeed, contrary to what had been claimed fifteen years before, when there was talk of certification under the licence agreement, BAe hastened to point out that the procedures could merely be resumed subsequent to a Romaero audit, but without the momentary interruption of the quality control exercised by the design authority making the process irreversible.[124] In fact, the negotiations continued, focusing specifically on obtaining a guarantee from the Romanian government to cover the costs of the programme. At the end of October 1993, BAe Airbus's managing director Chris Geoghegan explained to Ambassador Bache that the difficulty was solely of a financial and non-industrial nature, given the limited size of a programme for only eleven aircraft:

> BAe said that if the Romanian Government is willing, and has the finance, then BAe will have a role to play under their licence agreement. They will, however, require commercial terms. Those terms said BAe would be an undertaking on the outstanding debt, restarting of the technical assistance

[122] FOI/DTI: note for the file, 28.05.93; letter by Amb. Bache, 28.05.93; Rolls-Royce to George, 16.06.93; the head of ECGD Underwriting Division (Europe & Sub Saharan Africa) to Rolls-Royce, 16.07.93.

[123] FOI/FCO: note written on 17.07.93; minute by the Central European Department, 20.07.93; FOI/DTI: the 1st Secretary (Commercial) to DTI, [14.07.93]; Goddard, *The Rolls-Royce Tay Engine*, *op. cit.*, p. 97-100.

[124] FOI/FCO: note by the DTI European Division, 09.08.93; Bucharest to DTI, INTRA 19, 27.10.93.

> [...] and a considerable amount of design development in the UK for the Tay re-engining programme. BAe confirmed that Taiwan has nothing to do with the ROMBAC operation, and there is no competition between the ROMBAC 1-11 and the Taiwanese 146s. Especially as only 11 airframes are involved here. BAe are not convinced that Kiwi is a viable airline and has the money to buy the airframes.[125]

Even if Rolls-Royce and Lloyds were convinced that it would not be possible to find enough money without the guarantee that ultimately had to come from the Romanian government, it should be noted that BAe was talking about only eleven airplanes in all, implicitly excluding any continuation of the project beyond the original lot and, as Ambassador Bache emphasised to his interlocutors, also ruling out the opportunity to clearly distinguish between the construction of the new ROMBACs and the re-engining of TAROM and Romania's 1-11s. In June 1992, TAROM had announced that it wanted to upgrade its fleet by purchasing from only one supplier, Boeing, which was willing to set up a maintenance centre for aircraft in Romania and to contract out the construction of some components to Romaero. The contract was signed on August 6, 1993, while TAROM – though laboriously – succeeded in securing a long-term agreement with both Boeing and Airbus, from which it had previously bought three A310s.[126]

This commitment to the solvency and modernisation of TAROM was part of the plan to revive privatisations and, more generally, to carry forward the stabilisation of the economy being pursued at that time by Nicolae Văcăroiu's government. Would there still be enough funds to support the ROMBAC project as well? The Romanian government, in the meetings with Geoghegan, had promised a decision by the end of October, but this did not materialise. In a letter of November 4, the Secretary General of the Government, Viorel Hrebenciuc, authorised the managing director of Romaero, Tudorel Dumitraşcu, to continue negotiations with Lloyds and potential investors in order to determine the amount to be secured. On November 8, TAROM sent a letter of intent to Romaero for the re-engining of its 1-11 fleet. These were elements that in the past would have strengthened the negotiating power of Romaero, but not in this case. In mid-November, Goddard was anxious for confirmation of the fact that there were, as the Romanian counterparts claimed, legislative

125 FOI/DTI: note for the file, 01.11.93.

126 FOI/DTI: minute by the 1st Secretary (Commercial), 10.11.93; note for the file, 12.11.93; FOIA/DoS: Bucharest to SecState, No. 4825, 21.06.92; Bucharest to SecState, No. 7477, 09.08.93; Bucharest to SecState, No. 2353, 17.03.94.

impediments to securing loans for over three years (the procedures for reactivation of the ROMBAC project were expected to last four years). On November 30, 1993, Dumitrașcu finally urged a decision but the government's response was negative. This was indeed the end of the road for twenty-five years of Anglo-Romanian cooperation in the aeronautical industry – or at least it nullified the co-operation that had matured during an era now lost, marked by Détente between the blocs and neo-Keynesian policies.[127]

The final liquidation of the ROMBAC project – the remnants of the fuselage of aircraft No. 10, never finished, still lie rusting near Chitila, not far from Bucharest – and Romaero's reorientation plans were eventually realised, paradoxically, under the guidance of Dumitru Cocu, the engineer who had directed the production of the first ROMBAC 1-11 in 1980 and who became, from 1994, the general director of the company. Through a new contract with Boeing in March 1994 and an agreement with Florida West Aircraft Services (which sent seven Romanian expatriate mechanics to Bucharest to train Romaero's staff), the company became a sub-contractor of Boeing and a maintenance centre certified by the Federal Aviation Administration. This strategy, coupled with a cut in about two-thirds of the staff compared to 1990 levels, allowed Romaero to record profits from 1994 and to become an authentically export-oriented company, albeit owned by the state. While, until 1993, 80% of its activities were aimed at the domestic market, in 1997, 78% was destined for overseas commissions. The last testimony to the ambitious projects of a large Romanian aeronautical industry remained the old production line of the small Islander, of which more than 500 specimens were now being built.[128]

As for the ROMBAC, according to Sergiu Celac, who, as ambassador to London, pledged to support the revival of the project, the market studies on the issue were finally transmitted to Brazil, where they contributed to the success of the privatised Empresa Brasileira de Aeronáutica (EMBRAER) in the area of medium-range jet airliners.[129] Conversely, Cocu told an influential aeronautical magazine that for Romania '[...] to suggest that we can compete with Boeing and Airbus in promoting a

[127] FOI/FCO: the Secretary General of the Romanian Government to the Managing Director of Romaero, 04.11.93; FOI/DTI: Tarom to Romaero, 08.11.93; Goddard to the Commercial Counselor Bucharest, fax No. 312, 15.11.93.

[128] FOIA/DoS: Bucharest to SecState, No. 2008, 08.03.94; Bucharest to SecState, No. 10.644, 03.11.94; Bucharest to SecState, No. 12.093, 16.12.94; Toma Roman, „O ‚Casă a poporului' zburătoare", *Jurnalul Național*, 01.08.2009.

[129] Celac to the author, 29.03.2015.

100-seater is ludicrous'.[130] The pair of porcelain birds donated by Rolls-Royce to Ceauşescu during the 1978 United Kingdom state visit went to auction in 1999 along with other jewellery received by the dictator in his frequent contacts with foreign countries.[131]

130 Andrzej Jeziorski, "No Turning BAC", *Flight International*, 15.04.97, p. 27.
131 Kate Connolly, "Ceauşescu sale: Everything must go", *The Guardian*, 06.08.99.

References

Archival Sources

Archive of the Romanian Ministry of Foreign Affairs, Bucharest.

Archives du Ministère des affaires étrangères, La Courneuve.

Historical Archives of the European Communities, Fiesole.

Lloyds Banking Group Archives, London.

Papers obtained by the UK Department for Business, Innovation & Skills pursuant the Freedom of Information Act.

Papers obtained by the UK Foreign and Commonwealth Office pursuant the Freedom of Information Act.

Papers obtained from the U.S. Department of State pursuant the Freedom of Information Act.

Papers of Julian Amery, Churchill College, Cambridge.

Records of the U.K. Civil Aviation Administration, Gatwick.

The National Archives of Romania, Central Committee of the Romanian Communist Party, Bucharest.

The National Archives of the United Kingdom, Kew Gardens.

Published Sources

Akten zu Auswärtigen Politik der Bundesrepublik Deutschland, 1975, Band II, Munich, Oldenbourg, 2006.

Akten zu Auswärtigen Politik der Bundesrepublik Deutschland, 1976, Band I, Munich, Oldenbourg, 2007.

Akten zu Auswärtigen Politik der Bundesrepublik Deutschland, 1977, Band I, Munich, Oldenbourg, 2008.

Akten zu Auswärtigen Politik der Bundesrepublik Deutschland, 1977, Band II, Munich, Oldenbourg, 2008.

Akten zu Auswärtigen Politik der Bundesrepublik Deutschland 1978, Band I, Munich, Oldenbourg, 2009.

Anton, Mioara (ed.), *România şi Tratatul de la Varşovia. Conferinţele miniştrilor Afacerilor Externe şi ale adjuncţilor lor (1966-1991)*, Bucharest, Alpha, 2009.

Banu, Florian (ed.), *"Amorsarea" Revoluției. România anilor '80 văzută prin ochii Securității*, Târgoviște, Editura Cetatea de Scaun, 2012.

Barfield, Norman, "The BAC One-Eleven Family", *Aircraft Engineering*, Vol. 42, No. 3 (1970), p. 6-14.

Bennett, Gil – Hamilton, Keith (eds.), "Britain and the Soviet Union, 1968-72", in *Documents on British Foreign Policy Overseas*, Series III, Vol. 1, London, HMSO, 1997.

Bennett, Gil – Hamilton, Keith (eds.), "Détente in Europe, 1972-76", doc. 57, in *Documents on British Foreign Policy Overseas*, Series III, Vol. 3, London, Frank Cass, 2001.

"British Aircraft Corporation Three-Eleven", *Aircraft Engineering*, Vol. 40, No. 11 (1968), p. 22-23.

Budura, Romulus I. (ed.), *Relațiile Româno-Chineze 1880-1974: Documente*, Bucharest, Ministerul Afacerilor Externe – Arhivele Naționale, 2005.

Ceaușescu, Nicolae, *Dezvoltarea și modernizarea bazei tehnice-materiale a socialismului în România*, Bucharest, Ed. Politică, 1978.

Ceaușescu, Nicolae, *Raport la cel de-al XI-lea Congres al Partidului Comunist Român*, Bucharest, Ed. Politică, 1974.

Ceaușescu, Nicolae, *Raport la cel de-al XII-lea Congres al Partidului Comunist Român*, Bucharest, Ed. Politică, 1979.

Ceaușescu, Nicolae, *Raport la cel de-al XIII-lea Congres al Partidului Comunist Român*, Bucharest, Ed. Politică, 1984.

Ceaușescu, Nicolae, *Raport la Conferința Națională a Partidului Comunist Român, 16-18 decembrie 1982*, Bucharest, Ed. Politică, 1982.

Ceaușescu, Nicolae, *România pe drumul construirii societății socialiste multilateral dezvoltate*, Vol. 9, Bucharest, Ed. Politică, 1974.

Ceaușescu, Nicolae, *România pe drumul construirii societății socialiste multilateral dezvoltate*, Vol. 31, Bucharest, Ed. Politică, 1989.

Ceaușescu, Nicolae, *România pe drumul construirii societății socialiste multilateral dezvoltate*, Vol. 20, Bucharest, Ed. Politică, 1981.

Congresul al IX-lea al Partidului Comunist Român, 19-24 iulie 1965, Bucharest, Ed. Politică, 1965.

Cuvîntul tov. N. Ceaușescu la adunarea populației din capitală în piața palatului republicii, in *Principiile de bază ale politicii externe a României*, Bucharest, Ed. Politică, 1968.

Directivele Congresului al IX-lea al PRC cu privire la dezvoltarea economiei naționale în perioda 1966-1970, Bucharest, Ed. Politică, 1965.

Directivele Congresului al XI-lea al PCR cu privire la planul cincinal 1976-1980 și liniile directoare ale dezvoltării economico-sociale a României pentru perioada 1981-1990, Bucharest, Ed. Politică, 1974.

Directivele Congresului al X-lea al PCR cu privire la planul de dezvoltare economico-socială a României pe anii 1971-1975 și liniile directoare ale acesti dezvoltari pe perioada 1976-1980, in *Congresul al X-lea al Partidului Comunist Român, 6-10 august 1969*, Bucharest, Ed. Politică, 1969, p. 666-716.

Extinderea relațiilor economice, cooperarea în producție – o necesitate a progresului material, un factor al pacii și colaborarii internaționale, in *Congresul al X-lea al Partidului Comunist Român, 6-10 august 1969*, Bucharest, Ed. Politică, 1969, p. 42-44.

"Farnborough", *InterAvia Newsletter*, No. 10,584, 10.09.84, p. 2.

Ford, T.E., "The Britten-Norman Islander", *Aircraft Engineering*, Vol. 42, No. 8 (1970), p. 16-23.

Jackson, Marvin, *Romania's Economy at the End of the 1970s: Turning the Corner on Intensive Development*, in *East European Economic Assessment. A Compendium of Papers Submitted to the Joint Economic Committee, Congress of the United States*, part 1: *Country Studies, 1980*, Washington D.C., U.S. Government Printing Office, 1981, p. 232-298.

Let's Go with Labour for the New Britain. The Labour Party Manifesto for the 1964 General Elections, London, The Labour Party, 1964.

Mastny, Vojtech – Byrne, Malcolm (eds.), *A Cardboard Castle? An Inside History of the Warsaw Pact, 1955-1991*, Budapest, Central European University Press, 2005.

Năstase, Adrian (ed.), *Acordurile internaționale încheitate de România sau la care România a devenit parte (1990-1994)*, Bucharest, Metropol, 1995.

O politică de pace, colaborare și prietenie cu toate popoarele, Vol. 1, Bucharest, Ed. Politică, 1983.

Opriș, Petre (ed.), *Industria românească de apărare: documente, 1950-1989*, Ploiești, Ed. Universității Petrol-Gaze, 2007.

Preda Dumitru – Retegan Mihai (eds.), *1989 Principiul dominoului. Prăbușirea regimurilor comuniste europene. Documente*, Bucharest, Fundația Culturală Română, 2000.

Preda, Dumitru – Florian, Claudiu – Trasca, Ottmar (eds.), *România-Republica Federală Germană, (1966-1967)*. Vol. 1: *Începutul relațiilor diplomatice*, Bucharest, Ed. Enciclopedică, 2009.

Procacci, Giuliano (ed.), *The Cominform. Minutes of the Three Conferences (1947-1949)*, Milan, Feltrinelli, 1994.

Programul Partidului Comunist Român de făurire a societății socialiste multilateral dezvoltate și înaintare a României spre comunism, Bucharest, Ed. Politică, 1975.

Report of the Committee of Inquiry into the Aircraft Industry, Cmd. 2853, London, H.M. Stationary Office, 1965.

Revoluția Română văzută de ziariști americani și englezi, Bucharest, Editions Evenimentul, 1991.

Rolls-Royce and the RB.211 Aero-Engine, Cmd. 4860, London, HMSO, 1972.

"Rolls-Royce Engine Update", *Aircraft Engineering*, Vol. 48, No. 1 (1976), p. 5-8.

"Romania Signs for BAC One-Elevens and R-R Spey Engines", *Trade and Industry*, 23.06.78, p. 673.

"Romanians in Deal To Build 11 BAC 1-11s", *Financial Times*, 09.02.93.

"Rumanian Interest in Computers", *The Times*, 13.03.67

Sarsfield, Kate, "Romanian Reality", *Flight International*, 27.07.93, p. 33.

Sava, Constantin – Monac, Constantin (eds.), *Revoluția Română din decembrie 1989 retrăită prin documente si mărturii*, Bucharest, Axioma, 2001.

Savranskaya, Svetlana – Blanton, Thomas – Zubok, Vladislav (eds.), *Masterpieces of History. The Peaceful End of the Cold War in Europe, 1989*, Budapest – New York, CEU Press, 2010.

Stalin, Iosif Vissarionovich, *Economic Problems of Socialism in the USSR*, Moscow, Foreign Languages Publishing House, 1952.

Țăranu, Liviu (ed.), *Românii in „epoca de aur". Corespondență din anii 80*, Târgoviște, Editura Cetatea de Scaun, 2012.

"Tay-powered BAC One-Eleven Proposed", *Aircraft Engineering*, Vol. 55, No. 8 (1983), p. 14-16.

"The Civil Aircraft Industry", *Aircraft Engineering*, Vol. 42, No. 5 (1970), p. 5.

"The Future of the Industry", *Aircraft Engineering*, Vol. 39, No. 1 (1967), p. 7.

"The Industry on Plowden", *Aircraft Engineering*, Vol. 38, No. 2 (1966), p. 9.

"The Islander Saga", *Aircraft Engineering*, Vol. 42, No. 8 (1970), p. 6-14.

U.S. Congress, Office of Technological Assessment, *Technology and East-West Trade*, Washington D.C., GPO, 1979.

Valev, Emil, "Problems of Economic Development of the Danube Region of Romania, Bulgaria and the Soviet Union" (in Russian), *Вестника Московского университета*, No. 2 (1964), p. 56.

"VFW 614 – The Quietest Jet Transport in the World", *Aircraft Engineering*, Vol. 46, No. 6 (1974), p. 5-34.

Sources available online

- British Diplomatic Oral History Programme, at https://www.chu.cam.ac.uk/archives/collections/bdohp/.
- CIA Electronic Reading Room, at https://www.cia.gov/library/readingroom/home.

- Flight Magazine, past issues (1909-2004), at https://www.flightglobal.com/pdfarchive/index.html:
  - "Action Programme for European Aerospace", *Flight International*, 16.10.75, p. 553.
  - "And Now the Mainlander", *Flight International*, 14.12.72.
  - "BAe Restructures", *Flight International*, 07.01.89, p. 5.
  - "Britten-Norman Background", *Flight International*, 28.10.71, p. 679.
  - "Britten-Norman Move On", *Flight International*, 13.03.76, p. 624.
  - "Dowty To Acquire Rotol", *Flight International*, 26.12.58, p. 963.
  - Egozi, Arie, "IAR-99 Upgrade Flight-tested in Israel", *Flight International*, 28.02.94, p. 15.
  - "Egypt Buys Russian", *Flight International*, 23.03.72, p. 404.
  - "Fairey Britten-Norman Plans", *Flight International*, 28.12.72, p. 927.
  - "Fairey Buys Britten-Norman", *Flight International*, 07.09.72, p. 314.
  - "Fairey Call for Help", *Flight International*, p. 1172-1173.
  - "Features of the Tu-134", *Flight International*, 22.08.68, p. 288-290.
  - "FLS Favourite in One-Eleven Tay Project", *Flight International*, 25.05.93, p. 8.
  - Fulton, Kenneth, "Speys International", *Flight International*, 16.11.67, p. 805-811.
  - "Henry Coanda Visits Britain", *Flight International*, 28.01.71, p. 115.
  - "Industry International", *Flight International*, 21.11.63, p. 850.
  - "Iron Curtain Airlines News", *Flight International*, 11.02.55, p. 184-185.
  - Jeziorski, Andrzej, "No Turning BAC", *Flight International*, 15.04.97, p. 27.
  - Klapwijk Leaves VFW-Fokker, *Flight International*, 27.05.78, p. 1587.
  - Lambert, Mark, "Germany's Aerospace Industry", *Flight International*, 28.04.78, p. 1277.
  - "New Business-jet Agency", *Flight International*, 30.08.1980, p. 768.
  - "New Plans for Gosselies and Bembridge", *Flight International*, 19.04.73, p. 613.
  - "Paris: Biggest Ever", *Flight International*, 01.06.67, p. 899.
  - "Pilatus To Buy Britten-Norman", *Flight International*, 05.08.78.
  - Postlethwaite, Alan, "Coming in from the Cold", *Flight International*, 17.07.90, p. 26-28.
  - "Receiver for Britten-Norman", *Flight International*, 28.10.71, p. 670.
  - "Rolls-Royce Failure", *Flight International*, 11.02.71, p. 181-184.

- "Romania Looks for Western Investment", *Flight International*, 10.07.90, p. 13.
- "Romania Signs New One-Eleven Deal", *Flight International*, 17.07.90, p. 5.
- "Romanian One-Eleven Contract This Summer?", *Flight International*, 24.06.78, p. 1903.
- "Shorts To Take Over Islander and Trislander", *Flight International*, 27.08.77, p. 574.
- "Six Bids for Britten-Norman", *Flight International*, 11.11.71, p. 761-762.
- "SOKO Galeb and Jabstreb", *Flight International*, 28.11.68, p. 901-905.
- "The VFW-Fokker Divorce", *Flight International*, 01.03.80, p. 613.
- "The World Airlines", *Flight International*, 06.03.53, p. 313.
- "Three-Eleven Prospects", *Flight International*, 17.09.70, p. 431.
- "Tupolev Tu-154", *Flight International*, 18.11.71, p. 814-815.
- "Tupolev Tu-154/154A", *Flight International*, 24.10.74, p. 577.
- "World Airlines Survey", *Flight International*, 14.04.66, p. 637.

• Miscellaneous laws and decrees of Romania, at http://lege5.ro/:
- Decretul No. 155 din 19.06.1974 privind organizarea Consiliului Tehnic de Coordonare a Producției de Tehnică Militară.
- Decretul No. 162/1973 privind stabilirea normelor unitare de structura pentru unitatile economice în vigoare de la 11.05.1973.
- Decretul No. 419/1972 privind trecerea activității aeronautice civile şi a unităților de aviație civilă la Ministerul Forțelor Armate în vigoare de la 01.11.1972.
- Lege 1/1971 cu privire la activitatea de comerț exterior, de cooperare economică şi tehnico-ştiințifică a Republicii Socialiste România.
- Legea No. 12/1980 din 19 decembrie 1980 pentru întărirea autoconducerii muncitoreşti şi autogestiunii economico-financiare şi valutare în activitatea de comerț exterior şi cooperare economică internațională.
- Legea No. 15/1990 privind reorganizarea unităților economice de stat ca regii autonome şi societăți comerciale, în vigoare de la 07.09.1990.
- Legea No. 58/1991 privind privatizarea societăților comerciale şi vânzarea de active din patrimoniul acestora, în vigoare de la 16.08.1991.

• National Archives and Records Administration, access to archival databases, at https://aad.archives.gov/aad/.

Bibliography – General Political and Economic Matters

Aldcroft, Derek – Morewood, Steven, *Economic Change in Eastern Europe since 1918*, Aldershot, Elgar, 1995.

Autio-Sarasmo, Sari – Miklòssy, Katalin, "The Cold War from a New Perspective", in Sari Autio-Sarasmo and Katalin Miklòssy (eds.), *Reassessing Cold War Europe*, London, Routledge, 2011.

Bange, Oliver – Niedhart, Gottfried (eds.), *Helsinki 1975 and the Transformation of Europe*, New York, Berghahn, 2008.

Bell, Daniel, *The Coming of Post-Industrial Society. A Venture in Social Forecasting*, New York, Basic Books, 1973.

Berend, Ivan, *An Economic History of Twentieth-Century Europe. Economic Regimes from Laissez-Faire to Globalization*, Cambridge (MA), Cambridge University Press, 2006.

Berend, Ivan, *From the Soviet Bloc to the European Union. The Economic and Social Transformation of Central and Eastern Europe since 1973*, Cambridge, Cambridge University Press, 2009.

Bideleux, Robert – Jaffries, Ian, *A History of Eastern Europe: Crisis and Change*, London, Routledge, 1998.

Bideleux, Robert – Taylor, Richard (eds.), *European Integration and Disintegration: East and West*, London, Routledge, 1996.

Bischof, Günter – Karner, Stefan – Ruggenthaler, Peter (eds.), *The Prague Spring and the Warsaw Pact Invasion of Czechoslovakia in 1968*, Plymouth, Lexington Books, 2010.

Brenner, Robert, "The Economics of Global Turbolence", *New Left Review*, I/229 (1998), special issue.

Bronstone, Adam, *The European Bank for Reconstruction and Development. The Building of a Bank for East-Central Europe*, Manchester, Manchester University Press, 1999.

Brus, Wlodzimierz, "1966 to 1975: Normalisation and Conflict", in Michael Kaser (ed.), *The Economic History of Eastern Europe 1919-1975*, Vol. III: *Institutional Change within a Planned Economy*, Oxford, Clarendon, 1986, p. 222-228.

Burigana, David, "L'Europe, s'envolera-t-elle? Le lancement de Airbus et le sabordage d'une coopération aéronautique 'communautaire' (1965-78)", *Journal of European Integration History*, Vol. 13, No. 1 (2007), p. 91-109.

Cain, Frank, *Economic Statecraft During the Cold War. European Responses to the US Trade Embargo*, Abington, Routledge, 2013.

Carlin, Wendy – Landesmann, Michael, "From Theory into Practice? Restructuring and Dynamism in Transition Economy", *Oxford Review of Economic Policy*, Vol. 13, No. 2 (1997), p. 77-105.

Ceadel, Martin, "Cobden and Peace", in Anthony Howe – Simon Morgan (eds.), *Rethinking Nineteenth-Century Liberalism: Richard Cobden Bicentenary Essays*, Aldershot, Ashgate, 2006, p. 189-207.

Chassaigne, Philippe, *Les années 1970: fin d'un monde et origine de notre modernité*, Paris, Colin, 2008.

Eichengreen, Barry, *The European Economy Since 1945. Coordinated Capitalism and Beyond*, Princeton, Princeton University Press, 2007.

Eloranta, Jari – Olaja, Jari, *East-West Trade and the Cold War*, Jyväskylä, Jyväskylä University Press, 2005.

Eriksson, Sören, "Newly Industrialised Economies and the Aviation Industry: Economic and Industrial Policy Views in Asia", in Sören Eriksson and Harm-Jan Steenhuis (eds.), *The Global Commercial Aviation Industry*, Abingdon, Routledge, 2015, p. 231-296.

Godard, Simon, « Construire le bloc de l'Est par l'économie? La délicate émergence d'une solidarité internationale socialiste au sein du Conseil d'aide économique mutuelle », *Vingtième Siècle*, No. 109 (2011), p. 45-58.

Grynaviski, Eric, *Constructive Illusions. Misperceiving the Origins of International Cooperation*, Ithaca Cornell University Press, 2014.

Hall, Derek (ed.), *Tourism and Economic Development in Eastern Europe and the Soviet Union*, London, Belhaven Press, 1991.

Hanson, Philip, *The Rise and Fall of the Soviet Economy. An Economic History of the USSR from 1945*, New York, Longman, 2003.

Hayward, Keith, "European Union Policy and the European Aerospace Industry", *Journal of European Public Policy*, Vol. 1, No. 3 (1994), p. 347-365.

Hayward, Keith, *International Collaboration in Civil Aerospace*, London, Pinter, 1986.

James, Harold, *International Monetary Cooperation Since Bretton Woods*, Oxford, Oxford University Press, 1996.

Judt, Matthias, „Kompensationsgeschäfte der DDR. Instrumente einer europäischen Ost-West-Wirtschaftsintegration?", *Jahrbuch für Wirtschaftsgeschichte*, Vol. 49, No. 2 (2008), p. 117-138.

Kansikas, Suvi, *Socialist Countries Face the European Community. Soviet Bloc Controversies Over East-West Trade*, Brussels, P.I.E. Peter Lang, 2014.

Kornai, Jànos, *The Socialist System. The Political Economy of Communism*, Oxford, Clarendon, 1992.

Kotkin, Stephen, *Uncivil Society. 1989 and the Implosion of the Communist Establishment*, New York, The Modern Library, 2009.

Kramer, Mark, "The Czechoslovak Crisis and the Brezhnev Doctrine", in Carole Fink, Philipp Gassert and Detlef Junker (eds.), *1968: The World Transformed*, Cambridge, Cambridge University Press, 1998, p. 111-172.

Kramer, Mark, "The Prague Spring and the Soviet Invasion in Historical Perspective", in Günter Bischof, Stefan Karner and Peter Ruggenthaler (eds.), *The Prague Spring and the Warsaw Pact Invasion of Czechoslovakia in 1968*, Plymouth, Lexington Books, 2010, p. 33-58.

Levcik, Friedrich – Stankovsky, Jan, "East-West Economic Relations in the 1970s and 1980s", in Christopher Saunders (ed.), *East-West Trade and Finance in the World Economy. A New Look for the 1980s*, New York, St. Martin's Press, 1985, p. 77-124.

Lippert, Werner, *The Economic Diplomacy of Ostpolitik: Origins of NATO's Energy Dilemma*, New York, Berghahn, 2011.

Lipsey, Richard – Scarth, William (eds.), *Inflation and Unemployment: The Evolution of the Phillips Curve*, Cheltenham, Elgar, 2011.

Loth, Wilfried – Soutou, Georges-Henri, *The Making of Détente. Eastern and Western Europe in the Cold War, 1965-75*, Abingdon, Routledge, 2008.

Maier, Charles, "*'Malaise'. The Crisis of Capitalism in the 1970s*", in Niall Ferguson *et al.* (eds.), *The Shock of the Global. The 1970s in Perspective*, Cambridge (MA), Harvard University Press, 2010, p. 25-48.

Maier, Charles, *Dissolution. The Crisis of Communism and the End of East Germany*, Princeton, Princeton University Press, 1997.

Mastanduno, Michael, *Economic Containment. CoCom and the Politics of East-West Trade*, Ithaca, Cornell University Press, 1992.

McDaniel, Douglas, *United States Technology Export Control: An Assessment*, Westport, Praeger, 1993.

Morewood, Steven, "The Demise of Command Economies in the Soviet Union and its Outer Empire", Derek Aldcroft and Michael Oliver, *Economic Disasters of the Twentieth Century*, Chatenham, Elgar, 2007, p. 302-304.

Muller, Pierre, *Airbus: L'ambition européenne. Logique d'Etat, logique de marché*, Paris, L'Harmattan, 1989.

Nove, Alec – Hohmann, Hans Hermann – Seidenstecher, Gertrud (eds.), *The East European Economies in the 1970s*, London, Butterworths, 1982.

O'Donoghue, Jim – Goulding, Louise – Allen, Grahame, "Consumer Price Inflation Since 1750", *Economic Trends*, No. 604 (2004), p. 38-46.

Osterhammel, Jürgen, *Globalisation: A Short History*, Princeton, Princeton University Press, 2005.

Ouimet, Matthew, *The Rise and Fall of the Brezhnev Doctrine in Soviet Foreign Policy*, Chapel Hill, University of North Carolina Press, 2003.

Owen, Kenneth, *Concorde and the Americans. International politics of the Supersonic Transport*, London, Airlife Publishing, 1997.

Pötzl, Norbert, *Beitz. Eine deutsche Geschichte*, Munich, Heyne, 2011.

Romano, Angela – Romero, Federico, "European Socialist Regimes Facing Globalisation and European Co-operation: Dilemmas and Responses", *European Review of History*, Vol. 21, No. 2 (2014), p. 157-164.

Romano, Angela, *From Détente in Europe to European Détente*, Brussels, P.I.E. Peter Lang, 2009.

Romero, Federico – Pons, Silvio, "Europe Between the Superpowers, 1968-1981", in Antonio Varsori and Guia Migani (eds.), *Europe in the International Arena During the 1970s*, Brussels, P.I.E. Peter Lang, 2011, p. 85-97.

Romero, Federico, "Cold War Historiography at the Crossroads", *Cold War History*, Vol. 14, No. 4 (2014), p. 685-703.

Segreto, Luciano, "East-West Trade in Cold War Europe: National Interests and Hypocrisy", paper presented at the XIV International Economic History Congress, Session 111, Helsinki, August 2006.

Steiner, André, "The Council of Mutual Economic Assistance. An Example of Failed Economic Integration?", *Geschichte und Gesellschaft*, Vol. 39, No. 2 (2013), p. 240-258.

Steiner, André, "The Globalisation Process and the Eastern Bloc Countries in the 1970s and 1980s", *European Review of History*, Vol. 21, No. 2 (2014), p. 165-181.

Stone, David, "CMEA International Investment Bank and the Crisis of Developed Socialism", *Journal of Cold War Studies*, Vol. 10, No. 3 (2008), p. 48-77.

Stone, Randall, *Satellites and Commissars. Strategy and Conflict in the Politics of the Soviet-Block Trade*, Princeton, Princeton University Press, 1996.

Swain, Geoffrey – Swain Nigel, *Eastern Europe Since 1945*, Basingstoke, Palgrave, 2009.

Tomlinson, Jim, *Government and the Enterprise Since 1900. The Changing Problem of Efficiency*, Oxford, Clarendon, 1994.

Turnock, David, *The Economy of East-Central Europe, 1815-1989*, London, Routledge, 2004.

Van Brabant, Jozef, *Socialist Economic Integration*, Cambridge, Cambridge University Press, 1980.

van der Wee, Herman, *Prosperity and Upheaval. The World Economy, 1945-1980*, Berkeley, University of California Press, 1986.

Varsori, Antonio – Migani, Guia (eds.), *Europe in the International Arena During the 1970s*, Brussels, P.I.E. Peter Lang, 2011.

Weber, Steven, "Origins of the European Bank for Reconstruction and Development", *International Organization*, Vol. 48, No. 1 (1994), p. 1-38.

Wiener, Helgard – Slater, John, *East-West Technology Transfer. The Trade and Economic Aspects*, Paris, OECD, 1986.

Williams, Kieran, "The Russian View(s) of the Prague Spring", *Cold War History*, Vol. 14, No. 2 (2012), p. 128-141.

Zaccaria, Benedetto, *The EEC's Yugoslav Policy in Cold War Europe, 1968-1980*, Basingstoke, Palgrave, 2016.

Zloch-Christy, Iliana, *Debt problems of Eastern Europe*, Cambridge, Cambridge University Press, 1987.

Zurawicki, Leon, *Multinational Enterprises in the West and East*, Alphen aan den Rijn, Sijthoff & Nordhoff, 1979.

Bibliography – Britain: General, Political and Diplomatic Matters

Alexander, Philip, "The Tale of the Two Smiths: the Transformation of Commonwealth Policy, 1964-70", *Contemporary British History*, Vol. 20, No. 3 (2006), p. 303-321.

Benn, Anthony W., *Office Without Power. Diaries, 1968-72*, London, Arrow, 1989.

Benn, Anthony W., *Out of the Wilderness. Diaries, 1963-67*, London, Arrow, 1988.

Black, Lawrence – Pemberton, Hugh – Thane, Pat (eds.), *Reassessing 1970s Britain*, Manchester, Manchester University Press, 2012.

Bulpitt, Jim, "The Discipline of the New Democracy. Mrs. Thatcher's Domestic Statecraft", *Political Studies*, Vol. 34, No. 1 (1986), p. 19-39.

Campbell, John, *The Iron Lady*, London, Pimlico, 2004.

Coopey, Richard – Fielding, Steven – Tiratsoo, Nick (eds.), *The Wilson Governments 1964-1970*, London, Pinter, 1993.

Deighton, Anne, "*Ostpolitik* or *Westpolitik*? British Foreign Policy, 1968-75", *International Affairs*, Vol. 74, No. 4 (1998), p. 893-901.

Dockrill, Saki, "Defence and Détente: Britain, the Soviet Union, and the 1968 Czech Crisis", in Günter Bischof, Stefan Karner and Peter Ruggenthaler (eds.), *The Prague Spring and the Warsaw Pact Invasion of Czechoslovakia in 1968*, Plymouth, Lexington Books, 2010, p. 249-270.

Dockrill, Saki, *Britain's Retreat from East of Suez. The Choice Between Europe and the World?*, Basingstoke, Palgrave, 2002.

Ellis, Sylvia, *Britain, America, and the Vietnam War*, Westport, Praeger, 2004.

Engel, Jeffrey, *Cold War at 30,000 Feet. The Anglo-American Fight for Aviation Supremacy*, Cambridge (MA), Harvard University Press, 2007.

Evans, Eric, *Thatcher and Thatcherism*, London, Routledge, 2013.

Hamilton, Keith, *Transformational Diplomacy after the Cold War. Britain's Know-How Fund in Post-Communist Europe, 1989-2003*, Abingdon, Routledge, 2013.

Hughes, Geraint, "British Policy Towards Eastern Europe and the Impact of the Prague Spring: 1964-68", *Cold War History*, Vol. 4, No. 2 (2004), p. 115-139.

Hughes, Geraint, *Harold Wilson's Cold War. The Labour Government and East-West Politics*, 1964-1970, Woodbridge, Boydell&Brewer, 2009.

Jackson, Ben – Saunders, Robert (eds.), *Making Thatcher's Britain*, Cambridge, Cambridge University Press, 2012.

Ledger, Robert, "From Solidarity to 'Shock Therapy'. British Foreign Policy Towards Poland Under the Thatcher Government, 1980-1990", *Contemporary British History*, Vol. 30, No. 1 (2015), p. 99-118.

Leruez, Jacques, *Thatcher: la dame de fer*, Bruxelles, Andre Versaille, 2012.

O'Hara Glen – Parr, Helen, "The Rise and Fall of a Reputation", *Contemporary British History*, Vol. 20, No. 3 (2006), p. 295-302.

O'Hara, Glen, *Governing Post-War Britain: The Paradoxes of Progress, 1951-1973*, Basingstoke, Palgrave, 2012.

Ortolano, Guy, *The Two Cultures Controversy: Science, Literature and Cultural Politics in 1960s Britain*, Cambridge, Cambridge University Press, 2009.

Parr, Helen, *Britain's Policy Towards the European Community. Harold Wilson and Britain's World Role, 1964-1967*, London, Routledge, 2006.

Sharp, Paul, *Thatcher's Diplomacy. The Revival of British Foreign Policy*, Basingstoke, Macmillan, 1997.

Steward, Graham, *Bang! A History of Britain in the 1980s*, London, Atlantic Books, 2013.

Stoddart, Kristan, *Losing an Empire and Finding a Role. Britain, the USA, NATO and Nuclear Weapons, 1964-70*, Basingstoke, Palgrave, 2012.

Thatcher, Margaret, *The Downing Street Years*, London, Harper Collins, 1993.

Thorpe, Andrew, *A History of the British Labour Party*, Houndmills, Palgrave, 2008.

Vickers, Rhiannon, *Labour's Foreign Policy*, Manchester, Manchester University Press, 2011.

Vinen, Richard, "Thatcherism and the Cold War", in Ben Jackson and Robert Saunders (eds.), *Making Thatcher's Britain*, Cambridge, Cambridge University Press, 2012, p. 199-217.

Vinen, Richard, *Thatcher's Britain. The Politics and Social Upheaval of the 1980s*, London, Simon&Schuster, 2009.

White, Brian, *Britain, Détente and Changing East-West Relations*, London, Routledge, 1992.

Young John, *Twentieth Century Diplomacy: A Case Study of British Practice, 1963-1976*, Cambridge, Cambridge University Press, 2008.

Bibliography – Britain: Economics and Technology

Allen, George, *The British Disease. A Short Essay on the Nature and Causes of the Nation's Lagging Wealth*, London, The Institute of Economic Affairs, 1979.

Alt, James, "The Policy of Economic Decline in the 1970s", in Lawrence Black, Hugh Pemberton and Pat Thane (eds.), *Reassessing 1970s Britain*, Manchester, Manchester University Press, 2012, p. 25-40.

Alt, James, *The Politics of Economic Decline: Economic Management and Political Behaviour in Britain since 1964*, Cambridge, Cambridge University Press, 1979.

Bale, Tim, "Dynamics of a Non-Decision: the 'Failure' To Devaluate the Pound, 1964-67", *Twentieth Century British History*, Vol. 10, No. 2 (1999), p. 192-217.

Balogh, Thomas, *Planning for Progress. A Strategy for Labour*, London, Fabian Society, 1963.

Barnett, Correlli, *The Audit of War. The Illusion and Reality of Britain as a Great Nation*, Basingstoke, Macmillan, 1986.

Bar-Noi, Uri, "'Short List for a Long Haul': Britain's Role in the Process of Relaxing Strategic Export Controls During 1953-1954 Revisited", *Contemporary British History*, Vol. 26, No. 2 (2012), p. 173-194.

Bernstein, George, *The Myth of Decline: The Rise of Britain Since 1945*, London, Pimlico, 2004.

Booth, Alan – Glynn, Sean (eds.), *Modern Britain. An Economic and Social History*, London, Routledge, 1996.

Booth, Alan (ed.), *British Economic Development Since 1945*, Manchester, Manchester University Press, 1995.

Cairncross, Alec – Eichengreen, Barry, *Sterling in Decline. The Devaluations of 1931, 1949 and 1967*, Basingstoke, Palgrave, 1983.

Cairncross, Alec, *Managing the British Economy in the 1960s: A Treasury Perspective*, Basingstoke, Macmillan, 1996.

Cairncross, Alec, *The British Economy Since 1945. Economic Policy and Performance, 1945-1990*, Oxford, Blackwell, 1992.

Campbell-Kelly, Martin, *ICL. A Business and Technical History*, Oxford, Clarendon, 1989.

Cobham, David, *The Making of Monetary Policy in the UK, 1975-2000*, Chichester, Wiley, 2002.

Collins, Bruce – Robbins, Keith (eds.), *British Culture and Economic Decline*, London, Weidenfeld & Nicholson, 1990.

Edgerton, David, "The 'White Heat' Revisited. The British Government and Technology in the 1960s", *Twentieth Century British History*, Vol. 7, No. 1 (1996), p. 53-82.

Edgerton, David, *Warfare State. Britain, 1920-1970*, Cambridge, Cambridge University Press, 2006.

Gamble, Andrew, "Privatization, Thatcherism, and the British State", *Journal of Law and Society*, Vol. 16, No. 1 (1988), p. 1-20.

Goodhart, Charles, "The Economics of 'Big Bang'", in Ranald Michie (ed.), *The Development of London as a Financial Centre*, Vol. IV: *1945-2000*, London, Tauris, 2000, p. 261-277.

Harmon, Mark, *The British Labour Government and the 1976 IMF Crisis*, Basingstoke, Palgrave, 1997.

Hay, Colin, "Chronicles of a Death Foretold: the Winter of Discontent and Construction of the Crisis of British Keynesianism", *Parliamentary Affairs*, Vol. 63, No. 3 (2010), p. 446-470.

Henderson, David, "Two British Errors: Their Probable Size and Some Possible Reasons", *Oxford Economic Papers*, Vol. 29, No. 2 (1977), p. 159-205.

Hickson, Kevin, *The IMF Crisis of 1976 and British Politics*, London, Tauris, 2005.

Hill, Malcolm, *East-West Trade, Industrial Co-operation and Technological Transfer. The British Experience*, Aldershot, Gower, 1983.

Magazzino, Cosimo, *La politica economica di Margaret Thatcher*, Milano, Angeli, 2010.

Middleton, Roger, *Charlatans or Saviours? Economists and the British Economy from Marshall to Meade*, Northampton, Elgar, 1998.

Middleton, Roger, *The British Economy Since 1945. Engaging with the Debate*, Basingstoke, Macmillan, 2000.

Needham, Duncan, *UK Monetary Policy from Devaluation to Thatcher, 1967-1982*, Basingstoke, Palgrave, 2014.

Newton, Scott, "The Sterling Crises of 1964 and the Decision of Not To Devaluate", *Economic History Review*, Vol. 62, No. 1 (2009), p. 73-98.

Newton, Scott, "The Sterling Devaluation of 1967, the International Economy and Post-War Social Democracy", *English Historical Review*, No. 515 (2010), p. 912-945.

O'Hara, Glen, "'Dynamic, Exciting, Thrilling Change': The Wilson Government's Economic Policies, 1964-70", *Contemporary British History*, Vol. 20, No. 3 (2006), p. 383-402.

O'Hara, Glen, *From Dream to Disillusionment. Economic and Social Planning in 1960s Britain*, Basingstoke, Palgrave, 2006.

Pemberton, Hugh, "Relative Decline and British Economic Policy in the 1960s", *Historical Journal*, Vol. 47, No. 4 (2004), p. 989-1013.

Pemberton, Hugh, "Taxation and Labour's Modernisation Programme", *Contemporary British History*, Vol. 20, No. 3 (2006), p. 423-440.

Pollard, Sydney, *The Development of the British Economy, 1914-1960*, London, Arnold, 1962.

Pollard, Sydney, *The Wasting of the British Economy. British Economic Policy 1945 to the Present*, London, Croom Helm, 1982.

Radice, Hugo, "Experiences of East-West Industrial Cooperation: A Case Study of UK Firms in Electronics, Telecommunications and Precision Engineering Industries", in Christopher Saunders (ed.), *East-West Cooperation in Business: Inter-Firm Studies*, Springer, Vienna, 1977, p. 144-151.

Rogers, Chris, "Economic Policy and the Problem of Sterling under Harold Wilson and James Callaghan", *Contemporary British History*, Vol. 25, No. 3 (2011), p. 339-363.

Schenk, Catherine, *The Decline of Sterling. Managing the Retreat of an International Currency, 1945-1992*, Cambridge, Cambridge University Press, 2010.

Singh, Ajit, "The UK Industry and the World Economy. A Case of De-industrialisation", *Cambridge Journal of Economics*, Vol. I, No. 2 (1977), p. 113-136.

Slaven, Anthony, *British Shipbuilding, 1500-2010: A History*, Lancaster, Carnegie Publishing, 2013.

Snow, Charles, *The Two Cultures and the Scientific Revolution*, Cambridge, Cambridge University Press, 1959.

Supple, Barry, "Fear of Failing: Economic History and the Decline of Britain", in Peter Clarke and Clive Trebilcock (eds.), *Understanding Decline. Perceptions and Realties of British Economic Performance*, Cambridge, Cambridge University Press, 1997, p. 9-31.

Thompson, Noel, *Political Economy and the Labour Party*, Abington, Routledge, 2006.

Tomlinson, Jim, "Inventing 'Decline'. The Falling Behind of the British Economy in the Postwar Years", *Economic History Review*, Vol. 49, No. 4 (1996), p. 731-757.

Tomlinson, Jim, *The Labour Governments, 1964-1970.* Vol. III: *Economic Policy*, Manchester, Manchester University Press, 2003.

Tomlinson, Jim, "After Decline?", *Contemporary British History*, Vol. 23, No. 3 (2009), p. 395-406.

Tomlinson, Jim, "Thatcher, Monetarism and the Politics of Inflation", in Ben Jackson and Robert Saunders (eds.), *Making Thatcher's Britain*, Cambridge, Cambridge University Press, 2012, p. 62-77.

Tomlinson, Jim, "The Politics of Declinism", in Lawrence Black, Hugh Pemberton and Pat Thane (eds.), *Reassessing 1970s Britain*, Manchester, Manchester University Press, 2012, p. 41-60.

Tomlinson, Jim, "British Government and Popular Understanding of Inflation in the mid-1970s", *Economic History Review*, Vol. 67, No. 3 (2014), p. 750-768.

Tomlinson, Jim, "De-industrialisation Not Decline: A New Meta-narrative for Post-war British History", *Twentieth Century British History*, Vol. 27, No. 1 (2016), p. 76-99.

Wass, Douglas, *Decline to Fall. The Making of British Macro-Economic Policy and the 1976 IMF Crisis*, Oxford, Oxford University Press, 2008.

Winch, Donald, "A Great Deal of Ruin in a Nation", in Peter Clarke and Clive Trebilcock (eds.), *Understanding Decline. Perceptions and Realties of British Economic Performance*, Cambridge, Cambridge University Press, 1997, p. 32-48.

Whisler, Timothy, *The British Motor Industry, 1945-1994. A Case Study in Industrial Decline*, Oxford, Oxford University Press, 1999.

Wickham-Jones, Mark, *Economic Strategy and the Labour Party. Politics and Policy-Making, 1970-1983*, Basingstoke, Macmillan, 1996.

Wiener, Martin, *English Culture and the Decline of the Industrial Spirit, 1850-1980*, Cambridge, Cambridge University Press, 1981.

Bibliography – Britain: The Aviation Industry

Church, Robert, *The One-Eleven Story*, Tonbridge, Air-Britain Historians, 1996.

Edgerton, David, *England and the Aeroplane. An Essay on a Militant and Technological Nation*, Basingstoke, Macmillan, 1991.

Engel, Jeffrey, "'We are not concerned who the buyer is': Engines Sales and Anglo-American Security at the Dawn of the Jet Age", *History and Technology*, Vol. 17, No.1 (2000), p. 43-67.

Gardner, Charles, *British Aircraft Corporation*, London, Batsford, 1981.

Goddard, Ken, *The Rolls-Royce Tay Engine and the BAC One-Eleven*, RRHT Historical Series No. 30, Derby, 2000.

Gummett, Philip, "Civil and Military Aircraft in the UK", *History and Technology*, Vol. 9 (1992), p. 203-222.

Hartley, Michael, *The Rolls-Royce Spey*, RR Heritage Trust Technical Series No. 10, Derby, 2008.

Hayward, Keith, *Government and British Civil Aerospace. A Case Study in Post-War Technology Policy*, Manchester, Manchester University Press, 1983.

Hayward, Keith, *The British Aircraft Industry*, Manchester, Manchester University Press, 1989.

Hooker, Stanley, *Not Much of an Engineer. An Autobiography*, Marlborough, Airlife, 2002.

Johnman, Lewis – Lynch, Frances, "Technological Non-Cooperation: Britain and Airbus (1965-1969)", *Journal of European Integration History*, Vol. 12, No. 1 (2006), p. 125-140.

Johnman, Lewis – Lynch, Frances, "The Road to Concorde: Franco-British Relations and the Supersonic Project", *Contemporary European History*, Vol. 11, No. 2 (2002), p. 229-251.

Moore, Richard, "Bad Strategy and Bomber Dreams: A New View of the Blue Streak Cancellation", *Contemporary British History*, Vol. 27, No. 2 (2013), p. 145-166.

Pugh, Peter, *The Magic of a Name: The Rolls-Royce Story*, Vol. 2: *The Power Behind the Jets, 1945-1987*, Duxford, Icon, 2001.

Straw, Sean – Young, John, "The Wilson Government and the Demise of TSR-2, October 1964 – April 1965", *Journal of Strategic Studies*, Vol. 20, No. 4 (1997), p. 18-44.

Bibliography – Romania: General, Political and Diplomatic Matters

Andrei, Ştefan, *I se spunea Machiavelli. Ştefan Andrei în dialog cu Lavinia Betea*, Bucharest, Adevărul, 2011.

Anton, Mioara, *Ieşirea din cerc. Politica externă a regimului Gheorghiu-Dej*, Bucharest, INST, 2007.

Banu, Florian – Ţăranu, Liviu, *Aprilie 1964 „Primăvara de la Bucureşti". Cum s-a adoptat "Declaraţia de independenţă" a României?*, Bucharest, Ed. Enciclopedică, 2004.

Betea, Lavinia, *Alexandru Bârlădeanu despre Dej, Ceauşescu şi Iliescu. Convorbiri*, Bucharest, Ed. Evenimentul Românesc, 1997.

Betea, Lavinia, *Maurer şi lumea de ieri. Mărturii despre stalinizarea României*, Arad, Ed. Felix, 1995.

Bowd, Gavin, *La France et la Roumanie communiste*, Paris, L'Harmattan, 2008.

Bulei, Ion, *Brève histoire de la Roumanie*, Bucharest, Meronia, 2005.

Caroli, Giuliano, *La Romania nella politica estera italiana, 1919-1965. Luci e ombre di un'amicizia storica*, Milan, Negard, 2009.

Cătănuş, Dan, „Divergenţele româno-sovietice din CAER şi consecinţele lor asupra politicii externe a României", I, *Arhivele Totalitarismului*, No. 1-2 (2005), p. 68-80.

Chelaru, Rodica, *Culpe care nu se uită. Convorbiri cu Cornel Burtică*, Bucharest, Curtea Veche, 2001.

Connolly, Kate, "Ceausescu sale: Everything must go", *The Guardian*, 06.08.1999.

Deletant, Dennis, *Ceauşescu and the Securitate. Coercion and Dissent in Romania 1965-1989*, London, Hurst, 1995.

Deletant, Dennis, "Romania, 1948-1989. An Historical Overview", in *Romania and the Warsaw Pact: 1955-1989*, Cold War History International Project, Working Paper No. 43 (2004), p. 2-12.

Deletant, Dennis, "'Taunting the Bear': Romania and the Warsaw Pact, 1963-89", *Cold War History*, Vol. 7, No. 4 (2007), p. 495-507.

Dragomir, Elena, "The Perceived Threat of Hegemonism in Romania During the Second Détente", *Cold War History*, Vol. 12, No. 1 (2012), p. 111-134.

Dragomir, Elena, "Romania's participation in the Agricultural Conference in Moscow, 2-3 February 1960", *Cold War History*, Vol. 13, No. 3 (2013), p. 331-351.

Dragomir, Elena, "The Creation of the Council for Mutual Economic Assistance as Seen from the Romanian Archives", *Historical Research*, Vol. 88, No. 240 (2015), p. 355-379.

Dragomir, Elena, *Cold War Perceptions: Romania's Policy Change Towards the Soviet Union, 1960-1964*, Newcastle u. T., Cambridge Scholars Publishing, 2015.

Durandin, Catherine – Petre, Zoe, *Romania Since 1989*, New York, Columbia University Press, 2010.

Fischer, Mary, *Nicolae Ceauşescu. A Study in Political Leadership*, Boulder, Rienner, 1989.

Funderbunk, David, *Pinstripes and Reds. An American Ambassador Caught Between the State Department and the Romanian Communists, 1981-1985*, Washington, D.C., Selous Foundation Press, 1987.

Funderbunk, David, "Relations Between the United States and Romania During the First Half of the 1980s", in Paul Quinlan (ed.), *The United States and Romania. American-Romanian Relations in the Twentieth Century*, Woodland Hills, A.R.A., 1988, p. 147-159.

Gabanyi, Anneli Ute, *The Ceausescu Cult. Propaganda and Power Policy in Communist Romania*, Bucharest, Ed. Fudației Culturale Române, 2000.

Gallagher, Tom, *Modern Romania. The End of Communism, the Failure of Democratic Reform, and the Theft of a Nation*, New York, New York University Press, 2005.

Garthoff, Raymond, "When and Why Romania Distanced Itself from the Warsaw Pact", *CWIHP Bulletin*, No. 5 (1995), p. 111.

Gaston Marin, Gheorghe, *În serviciul României lui Gheorghiu-Dej. Însemnări din viață*, Bucharest, Evenimentul Românesc, 2000.

Gheorghe, Eliza, *The Romanian Intelligence Services During the Cold War: How Small Powers Can Sometimes Be Strong*, M.A. dissertation, Georgetown University, 2010.

Gheorghe, Eliza, "Atomic Maverick: Romania's Negotiations for Nuclear Technology, 1964-1970", *Cold War History*, Vol. 13, No. 3 (2013), p. 373-392.

Gheorghe, Eliza, "Nicolae Ceauşescu", in Steven Casey and Jonathan Wright (eds.), *Mental Maps in the Era of Détente and the End of the Cold War 1968-91*, Basingstoke, Palgrave, 2015, p. 60-80.

Gilberg, Trond, *Nationalism and Communism in Romania. The Rise and Fall of Ceauşescu's Personal Dictatorship*, Boulder, Westview, 1990.

Gridan, Irina, « Parallélismes et convergences (III). Ion Gheorghe Maurer a Paris a l'été 1964: Realpolitik et offensive du sourire », *Arhivele Totalitarismului*, Vol. XVII, No. 64-65 (2009), p. 99-122.

Gridan, Irina, « Du communisme national au national-communisme: réactions a la soviétisation dans la Roumanie des années 1960 », *Vingtième Siècle*, No. 109 (2011), p. 113-127.

Grosescu, Raluca, *Les communistes dans l'après-communisme. Trajectoires de conversion politique de la nomenklatura roumaine après 1989*, Paris, Houdiard, 2012.

Harrington, Joseph – Courtney, Bruce, *Tweaking the Nose of the Russians. Fifty years of American-Romanian Relations, 1940-1990*, Boulder, East European Monographs, 1991.

Herta, Laura-Maria, *De la relațiile româno-sârbe la relațiile româno-iugoslave. Interpretări constructiviste*, Cluj, Ed. Fund. Studii Europee, 2012.

Ionescu, Mihai, "Introduction", in *Romania and the Warsaw Pact: 1955-1989*, Cold War History International Project, Working Paper No. 43 (2004), p. 2-12.

Ionescu, Mihai, "Romania, Ostpolitik and the CSCE, 1967-1975", in Oliver Bange and Gottfried Niedhart (eds.), *Helsinki 1975 and the Transformation of Europe*, New York, Berghahn, 2008, p. 129-143.

Ionescu-Gură, Nicoleta, "Studiu introductiv", in Florica Dobre (ed.), *Membrii C.C. al P.C.R. 1945-1989. Dicționar*, Bucharest, Ed. Enciclopedică, 2004, p. 5-56.

Ivan, Ruxandra, *La politique étrangère roumaine (1990-2006)*, Brussels, Éditions de l'Université de Bruxelles, 2009.

Klingman, Gail, *The Politics of Duplicity. Controlling Reproduction in Ceausescu's Romania*, Berkeley, University of California Press, 1998.

Linden, Ronald, "After the Revolution: A Foreign Policy Bound to Change", in Daniel Nelson (ed.), *Romania after Tyranny*, Boulder, Westview Press, 1992, p. 203-238.

Mănescu, Corneliu, *Convorbiri nedeterminate. Corneliu Mănescu în dialog cu Lavinia Betea*, Bucharest, Polirom, 2001.

Moraru, Constantin, *Politica externă a României, 1958-1964*, Bucharest, Ed. Enciclopedică, 2009.

Munteanu, Mircea, "When the Levee Breaks. The Impact of the Sino-Soviet Split and the Invasion of Czechoslovakia on Romanian-Soviet Relations, 1967-1970", *Journal of Cold War Studies*, Vol. 12, No. 1 (2010), p. 43-61.

Neagoe-Pleşa, Elis, „1968 – anul reformării agenturii Securităţii", *Caietele CNSAS*, Vol. 1, No. 1 (2008), p. 9-22.

Nicolaescu, Sergiu, *Cartea Revoluţiei Române din Decembrie 1989*, Bucharest, Ed. Ion Cristoiu, 1999.

Niculescu-Mizil, Paul, *O istorie trăită*, Vol. 1, Bucharest, Ed. Enciclopedică, 1997.

Olteanu, Costantin, „Din relaţiile militare româno-iugoslave (1980-1985)", *Opinia Naţionala*, No. 550 (24.10.2011), p. 1,4.

Opriş, Petre, *Criza poloneză de la începutul anilor 80. Reacţia conducerii Partidului Communist Român*, Ploieşti, Petroleum-Gas University Publishing House, 2008.

Pacepa, Ion M., *Red Horizons. Chronicles of a Communist Spy Chief*, Washington, Gateway, 1987.

Păun, Ştefan – Lungu, Corneliu M. – Oşca, Alexandu, *Revoluţia Română din 1989 la Bucureşti*, Bucharest, Ed. IRRD, 2015.

Phinnemore, David, "Romania and Euro-Atlantic Integration Since 1989: A Decade of Frustration?", in Duncan Light and David Phinnemore (eds.), *Post-communist Romania. Coming to Terms with Transition*, Basingstoke, Palgrave, 2001, p. 245-269.

Retegan, Mihai, *In the Shadow of Prague Spring: Romania Foreign Policy and the Crisis in Czechoslovakia, 1968*, Iaşi, Centre for Romanian Studies, 2000.

Săndulescu, Şerban, *Lovitura de stat a confiscat revoluţia română*, Bucharest, Omega Press, 1996.

Scurtu, Ioan, *Revoluţia Română din decembrie 1989 în context internaţional*, Bucharest, Ed. Redacţiei Publicaţiilor pentru Străinătate, 2009.

Scutaru, Beatrice, *Les relations entre les sociétés française et roumaine des années 1960 à 1995: un atout pour l'ancrage de la Roumanie à l'Europe?*, Ph.D. dissertation, Université d'Angers, 2013.

Siani-Davies, Peter, *The Romanian Revolution of December 1989*, Ithaca, Cornell University Press, 2005.

Stan, Valentin, *România şi eşecul campaniei Vest*, Bucharest, Ed. Universităţii din Bucareşti, 1999.

Tismăneanu Vladimir – Vasile, Cristian, *Perfectul acrobat. Leonte Răutu. Maştile răului*, Bucharest, Humanitas, 2008.

Tismăneanu, Vladimir, *Gheorghiu-Dej and the Romanian Workers' Party: From the De-Sovietization to the Emergence of National Communism*, Cold War History International Project, Working Paper No. 37 (2002).

Tismăneanu, Vladimir, *Stalinism for All Seasons. A Political History of Romanian Communism*, Berkeley, University of California Press, 2003.

Verona, Sergiu, *Military Occupation and Diplomacy. Soviet Troops in Romania: 1944-1958*, London, Duke University Press, 1992.

Weiner, Robert, "The U.S. Policy of Differentiation Toward Romania", in Paul Quinlan (ed.), *The United States and Romania. American-Romanian Relations in the Twentieth Century*, Woodland Hills, A.R.A., 1988, p. 129-145.

Williams, Kieran, "Romania's Resistance to the USSR", in Kristen Williams, Steven Lobell and Neal Jesse (eds.), *Beyond Great Powers and Hegemons. Why Secondary States Support, Follow, or Challenge*, Stanford, Stanford University Press, 2012, p. 33-41.

Bibliography – Romania: Economics and Technology

Banu, Florian, „Capitaliştii avant la lettre: Securitatea şi operaţiunile valutare speciale din anii 80", *Caietele CNSAS*, Vol. IV, No. 1-2 (2011), p. 109-136.

Banu, Florian, *Asalt asupra economiei României, de la Solagra la SOVROM (1936-1956)*, Bucharest, Nemira, 2004.

Bălan, Ştefan – Mihăilescu, Nicolae, *Istoria ştiinţei şi tehnicii în România. Date cronologice*, Bucharest, Ed. Academiei RSR, 1983.

Ben-Ner, Avner – Montias, John, "Economic Systems Reforms and Privatization in Romania", in Saul Estrin (ed.), *Privatization in Central and Eastern Europe*, London, Longman, 1994, p. 279-310.

Costantinescu, Nicolae (ed.), *Istoria economică a României*, Vol. II, Bucharest, Ed. Economică, 2000.

Gabanyi, Anneli Ute, "The New Business Elite: From Nomenklatura to Oligarchy", in Henry Carey (ed.), *Romania Since 1989: Politics, Economics, and Society*, Lanham, Lexington Books, 2004, p. 353-372.

Gheorghe, Eliza, "Building Détente in Europe? East-West Trade and the Beginnings of Romania's Nuclear Programme, 1964-70", *European Review of History*, Vol. 21, No. 2 (2014), p. 235-253.

Kaser, Michael – Spigler, Iancu, "Economic Reform in Romania", in Alec Nove, Hans-Hermann Hohmann and Gertrud Seidenstecher (eds.), *The East European Economies in the 1970s*, London, Butterworths, 1982, p. 253-279.

Kotkin, Stephen, *The Kiss of Debt. The East Bloc Goes Borrowing* in Niall Ferguson *et al.* (cds.), *The Shock of the Global. The 1970s in Perspective*, Cambridge (MA), Harvard University Press, 2010, p. 80-93.

Linden, Ronald, *Communist States and International Change. Romania and Yugoslavia in Comparative Perspective*, Boston, Allen & Unwin, 1987.

Montias, John, *Economic Development in Communist Romania*, Cambridge (MA), MIT Press, 1967.

Montias, John, "Romania's Foreign Trade Between East and West", in Paul Marer and John Montias (eds.), *East European Integration and East-West Trade*, Bloomington, Indiana University Press, 1980, p. 326-329.

Mureşan, Maria, "Romania's Integration in Comecon. The Analysis of a Failure", *The Romanian Economic Journal*, Vol. 9, No. 30 (2008), p. 27-58.

Murgescu, Bogdan, *România şi Europa. Acumularea decalajelor economice (1500-2010)*, Iaşi, Polirom, 2010.

Negrescu, Dragoş, *A Decade of Privatization in Romania*, Bucharest, Romanian Centre for Economic Policy, 1999.

Nicolaescu, Teodor, "Privatization in Romania. The Case for Financial Institutions", in Donald Fair and Robert Raymond (eds.), *The New Europe: Evolving Economic and Financial Systems in East and West*, Dordrecht, Springer, 1993, p. 101-109.

Rotariu, Ilie, *Globalizare şi turism: cazul României*, Sibiu, Continent, 2004.

Shen, Raphael, *The Restructuring of Romania's Economy. A Paradigm of Flexibility*, London, Praeger, 1997.

Smith, Alan, "The Transition to a Market Economy in Romania and the Competitiveness of Exports", in Iancu Spigler, *Economic Reform in Romanian Industry*, Oxford, Oxford University Press, 1973.

Stanciu, Cezar, "Crisis Management in the Communist Bloc: Romania's Policy Towards the USSR in the Aftermath of the Prague Spring", *Cold War History*, Vol. 13, No. 3 (2013), p. 353-372.

Sută, Nicolae (ed.), *Istoria comerţului exterior românesc. O prezentare sintetică*, Bucharest, Eficient, 1998.

Ţăranu, Liviu, *România in Consiliul de Ajutor Economic Reciproc, 1949-1965*, Bucharest, Ed. Enciclopedică, 2007.

Tsantis, Andreas – Pepper, Roy, *Romania: The Industrialization of an Agrarian Economy Under Socialist Planning*, Washington D.C., The World Bank, 1979.

Turnock, David, *The Romanian Economy in the Twentieth Century*, London, Croom Helm, 1986.

Turnock, David, "Romania", in Derek Hall (ed.), *Tourism and Economic Development in Eastern Europe and the Soviet Union*, London, Belhaven Press, 1991, p. 203-219.

Turnock, David, *Restructuring the Romanian Economy: A Study of Transition 1990-2006*, Cheltenham, Elgar, 2007.

Turnock, David, *Aspects of Independent Romania's Economic History with Particular Reference to Transition for EU Accession*, Farnham, Ashgate, 2013.

Bibliography – Romania: The Aviation Industry

Dumitraşcu, Ion, *Industria Aeronăutică Română Braşov (1925-1945)*, Bucharest, Domino, 2008.

Ispaş, Ştefan, *Romanian Wings. A Review of Romania's Contributions in Aviation Development through more than 60 Years*, Bucharest, IAR, n.d.

Opriş, Petre, „Colaborarea lui Ceauşescu cu British Aerospace", *Jurnalul Naţional*, 17.08.09.

Opriş, Petre, „Avioane de vânătoare Mig-23 şi Mig-29, pentru forţele aeriene române (1977-1989)", *Art-Emis Academy*, 11.03.2012, at http://www.art-emis.ro/analize/875-avioane-de-vanatoare-mig-23-mig-29-pentru-forele-aeriene-romane-1977-1989.html.

Opriş, Petre, „Avioane civile utilizate de şefii statului român (1957-1991)", *Romanian Journal of History and International Studies*, Vol. 1, No. 2 (2014), p. 241-254.

Opriş, Petre, „Tatonări ale regimului de la Bucureşti privind achiziţia unor avioane ‚Boeing', 1964-1974", *Arhivele Totalitarismului*, Vol. XXII, No. 82-83 (2014), 81-92.

Roman, Toma, „O ‚Casă a poporului' zburătoare", *Jurnalul Naţional*, 01.08.09.

Tătaru, Simion, *31 octombrie 2004. 30 de ani de la primul zbor al avionului de luptă si sprijin aerian IAR-93*, paper sent to the Author on 28.01.2015.

Tătaru, Simion – Olivotto, Constantin – Stefan, Vasile, "IAR-XT/Next Generation", *INCAS Bulletin*, Vol. 4, No. 1 (2012), p. 125-132.

Turturică, Sorin (ed.), *„Pe aviatori, lăsaţi-i să zboare!" Un dialog între general av. (r.) Aurel Niculescu şi Sorin Turturică*, Bucharest, Ed. Anima, 2008.

Ţăranu, Liviu, „Afacerea Fokker", *Magazin Istoric*, No. 487 (2007), p. 49-52.

Vârlan, Manuela – Vârlan, Doru, *Şcoala superioară de aeronautică civilă*, Pleseşti, Asociaţia aerospaţială română, 2010.

Index

A

B

C

D

E

F

G

H

I

J

K

L

M

N

O

P

Q

R

S

EUROCLIO – Published Books

N° 101 – *Experts and Expertise in Science and Technology in Europe since the 1960s Organized civil Society, Democracy and Political Decision-making*. Christine Bouneau, David Burigana (eds.), 2018.

N° 99 – *The History of the European Monetary Union. Comparing Strategies amidst Prospects for Integration and National Resistance.* Daniela Preda, 2017.

N° 98 – *The European Union Approach Towards Western Sahara.* Marco Balboni and Giuliana Laschi, 2016.

N° 97 – *Europe between Imperial Decline and Quest for Integration. Pro-European Groups and the French, Belgian and British Empires (1947-1957).* Laura Kottos, 2016.

N° 96 – *Réinventer la diplomatie. Sociabilités, réseaux et pratiques diplomatiques en Europe depuis 1919 / Reshaping Diplomacy. Networks, Practices and Dynamics of Socialization in European Diplomacy since 1919.* Vincent Genin, Matthieu Osmont and Thomas Raineau (eds.), 2016.

N° 95 – *Sometimes Speaking with a Single Voice. The European Community as an International Actor, 1969-1979.* Lorenzo Ferrari, 2016.

N° 94 – *Competition Law Challenges in the Next Decade.* Sofia Oliveira Pais (ed.), 2016.

N° 93 – *The Borders of Schengen.* Alice Cunha, Marta Silva, Rui Frederico (eds.), 2015.

N° 92 – *Que dire à l'Espagne ? De l'isolement franquiste à la démocratie européiste, la France au défi, 1957-1979.* Pablo Martin-Pañeda, 2015.

N° 91 – *Francis Delaisi, du dreyfusisme à « l'Europe nouvelle ».* Éric Bussière, Olivier Dard, Geneviève Duchenne (dir.), 2015.

N° 90 – *European Parties and the European Integration Process, 1945-1992.* Lucia Bonfreschi, Giovanni Orsina, Antonio Varsori (eds.), 2015

N° 89 – *Européanisation et démocratisation des États baltes dans la période de préadhésion à l'UE. Le rôle de la conditionnalité politique des organisations européennes.* Anne-Sylvie Pigeonnier, 2015.

N° 88 – *Le Continent basket. L'Europe et le basket-ball au XX*[e] *siècle.* Fabien Archambault, Loïc Artiaga, Gérard Bosc (dir.), 2015.

N° 87 – *The Convention on the Future of Europe. How States Behave in a New Institutional Context of Negotiation.* Francesco Marchi, 2015.

N° 86 – *Christian Democrat Internationalism. Its Action in Europe and Worldwide from post World War II until the 1990s. Volume III. The European People's Party Continental and Social Cooperation.* Jean-Dominique Durand (ed.), 2014.

N° 85 – *Pela Paz! For Peace! Pour la paix ! (1849-1939)*. Maria Manuela Tavares Ribeiro, Maria Fernanda Rollo, Isabel Maria Freitas Valente, Alice Cunha (eds.), 2014.

N° 84 – *Economic Crisis and New Nationalisms. German Political Economy as Perceived by European Partners*. Antonio Varsori, Monika Poettinger (eds.), 2014.

N° 83 – *European Constitutionalism. Historical and Contemporary Perspectives*. Alexandre Dupeyrix, Gérard Raulet (eds.), 2014.

N° 82 – *Cross-Border Cooperation Structures in Europe. Learning from the Past, Looking to the Future*. Luis Domínguez, Iva Pires (eds.), 2014.

N° 81 – *The European Community and the World. A Historical Perspective*. Giuliana Laschi (ed.), 2014.

N° 80 – *Christian Democrat Internationalism. Its Action in Europe and Worldwide from Post World War II until the 1990s. Volume II: The Development (1945-1979). The Role of Parties, Movements, People*. Jean-Dominique Durand (ed.), 2013.

N° 79 – *Christian Democrat Internationalism. Its Action in Europe and Worldwide from Post World War II until the 1990s. Volume I: The Origins*. Jean-Dominique Durand (ed.), 2013.

N° 78 – *Investment Banking History. National and Comparative Issues (19th-21st centuries)*. Hubert Bonin & Carlo Brambilla (eds.), 2013.

N° 77 – *Citizenship and Solidarity in the European Union. From the Charter of Fundamental Rights to the Crisis, the State of the Art*. Mariana Canotilho, Pedro Madeira Froufe & Alessandra Silveira (eds.), 2013.

N° 76– *The OSCE: Soft Security for a Hard World. Competing Theories for Understanding the OSCE*. Roberto Dominguez (ed.), 2013.

N° 75– *Euroscepticism, Europhobia and Eurocriticism. The Radical Parties of the Right and Left* vis-à-vis *the European Union*. Cesáreo Rodríguez-Aguilera de Prat, 2013.

N° 74 – *Fermer les mines en construisant l'Europe. Une histoire sociale de l'intégration européenne*. Nicolas Verschueren, 2012

N° 73 – *Reflets de la construction européenne. Réflexions, références et refus du débat sur l'Europe*. Bertrand Vayssière (dir.), 2012.

N° 72 – *The International Relations of Local Authorities. From Institutional Twinning to the Committee of the Regions: Fifty Years of European Integration History*. Fabio Zucca, 2012.

N° 71 – *Histoire du Conseil de l'Europe (1949-2009)*. Birte Wassenberg, 2012.

N° 70 – *The EU, Security and Transatlantic Relations*. Finn Laursen (ed.), 2012.

Euroclio European Network

together with the SEGEI Network

Coordination : Chaire Jean Monnet d'histoire de l'Europe contemporaine (Cehec)
Collège Erasme, 1, place Blaise-Pascal, B-1348 Louvain-la-Neuve

Germany

Jürgen Elvert
Wilfried Loth

Belgium

Julie Cailleau
Jocelyne Collonval
Yves Conrad
Gaëlle Courtois
Pascal Deloge
Geneviève Duchenne
Vincent Dujardin
Michel Dumoulin
Roch Hannecart
Pierre-Yves Plasman
Béatrice Roeh
Corine Schröder
Caroline Suzor
Pierre Tilly
Arthe Van Laer
Jérôme Wilson
Natacha Wittorski

Spain

Enrique Moradiellos
Mercedes Samaniego Boneu

France

Françoise Berger
Marie-Thérèse Bitsch
Gérard Bossuat
Éric Bussière
Jean-François Eck
Catherine Horel
Philippe Mioche
Marine Moguen-Toursel
Sylvain Schirmann
Matthieu Trouvé
Laurent Warlouzet
Emilie Willaert

Hungary

Gergely Fejérdy

Italy

David Burigana
Elena Calandri
Eleonora Guasconi
Luciano Segreto
Antonio Varsori

Luxemburg

Charles Barthel
Etienne Deschamps
Jean-Marie Kreins
René Leboutte
Robert Philippart
Corine Schröder
Gilbert Trausch

The Netherlands

Anjo Harryvan
Jan W. Brouwer
Jan van der Herst

Poland

Józef Laptos
Zdzisiaw Mach

Switzerland

Antoine Fleury
Lubor Jilek